# THE SCIENCE OF GYMNASTICS

*The Science of Gymnastics* is a comprehensive and accessible introduction to the fundamental physiological, biomechanical and psychological principles underpinning this most demanding of sports.

Drawing on cutting-edge scientific research, and including contributions from leading international sport scientists and experienced coaches, the book represents an important link between theory and performance. With useful summaries, data and review questions included throughout, the book examines every key aspect of gymnastic training and performance, including:

- energetic, physical and physiological assessment
- training principles
- diet, nutrition and supplementation
- growth and development issues
- kinetics and kinematics
- angular and linear motion
- angular momentum
- stress, anxiety and coping
- motivation and goal setting
- mental skills training for practice and competition
- the psychology of learning and performance.

In a concluding section the authors reflect on how fundamental scientific components (physiology, biomechanics and psychology) interact to enhance gymnastic performance, helping students to develop a better understanding of the relationship between sport science and sporting performance. *The Science of Gymnastics* is essential reading for all students, coaches and researchers with an interest in gymnastics or applied sport science.

**Monèm Jemni** is a Principal Lecturer in Sport and Exercise Science at the University of Greenwich, UK. A former international gymnast and coach, he has provided sport science support to many teams and competitors in gymnastics, including members of the French national team. He is widely recognized as a leading expert in the bioenergetics, physiology and health-related issues of gymnastics, as well as the enhancement of gymnastics performance through training.

# THE SCIENCE OF GYMNASTICS

*Edited by Monèm Jemni*

WITH WILLIAM A. SANDS, JOHN H. SALMELA,
PATRICE HOLVOET AND MARIA GATEVA

LONDON AND NEW YORK

First published 2011
by Routledge
2 Park Square, Milton Park, Abingdon, Oxon, OX14 4RN

Simultaneously published in the USA and Canada
by Routledge
711 Third Avenue, New York, NY 10017

*Routledge is an imprint of the Taylor & Francis Group, an informa business*



Typeset in Goudy by Glyph International



*British Library Cataloguing in Publication Data*
A catalogue record for this book is available from the British Library

*Library of Congress Cataloging in Publication Data*
The Science of Gymnastics
Edited by Monèm Jemni
p. cm.
1. Gymnastics–Physiological aspects. 2. Human mechanics.
I. Jemni, Monèm.
RC1220.G95S35 2011
612.7'6–dc22 2010022060

ISBN 13: 978-0-415-54990-5 (hbk)
ISBN 13: 978-0-415-54991-2 (pbk)
ISBN 13: 978-0-203-87463-9 (ebk)

# DEDICATIONS

This book is dedicated:

- to my parents for their everlasting benediction in each step of my life;
- to my brothers and sister for their endless support;
- to all the teachers who crossed my road;
- to all gymnasts, students, coaches and scientists who are continuously making this sport a real joy;
- to Bill, John, Patrice and Maria, for making this book become a reality.

Monèm Jemni

To Natalie and Ed,
I would thank you from the bottom of my heart, but for you my heart has no bottom.

Bill

My contribution to this book is dedicated to Boris Shaklin, who won six Olympic gold medals from 1956 to 1964, whom I met in 2004 in Kiev, and who was an inspiration to me as a young gymnast. He was the first documented gymnast to perform mental training and mental practice before each event, decades before the academic world ever heard of these terms.

John

To my work and professional colleagues, to my friends who are gymnastic coaches, to my students and to my gymnasts, whose questions and comments have inspired my research and my contribution.

Patrice

My chapter in this book is dedicated to the persons who supported me when I was a gymnast. I also dedicate it to all the professionals working under the great umbrella of gymnastics.

Maria

# CONTENTS

# FIGURES AND TABLES

## Figures

## Tables

# ABOUT THE AUTHORS

## Editor

**Dr Monèm Jemni, Ph.D.**
University of Greenwich
London, UK

Monèm is a former gymnast and international coach, and is now a specialised researcher in the physiology of gymnastics, with a world leading reputation as a result of his cutting-edge investigations. He has written and edited more than 100 papers, and has lectured, and received outstanding academic awards in universities in France, Tunisia, the USA and currently, the UK.

## Contributors

**Dr William A. Sands, Ph.D., FACSM**
Mesa State College
Colorado, USA

William is a former gymnast and World Championship coach. He has held academic positions with the US Olympic Committee and several Universities and has served as Director of Research and Development for USA Gymnastics and Chair of the US Elite Coaches Association for Women's Gymnastics. He has written and edited 17 books and over 200 articles.

**Prof. John H. Salmela, Ph.D.**
University of Ottawa
Ottawa, Canada

John is currently retired after 29 years of working at the Universities of Laval, Montreal and Ottawa. A former gymnast, he has worked as an international coach and sports psychologist, and has written and edited more than 20 books and 250 articles. He acted as research chairman and sports psychologist for the Canadian men's national gymnastics team for 19 years.

**Dr Patrice Holvoet, Ph.D.**
University of Lille 2
Lille, France

Patrice is a former gymnast and coach and is an accredited physical education teacher. His research in the biomechanics of gymnastics has contributed to the development of the sport in France and around the world. He has written and edited more than 30 articles and manuscripts in French and English.

**Dr Maria Gateva, Ph.D.**
National Sports Academy
Sofia, Bulgaria

Maria is a former member of the Bulgarian National Rhythmic Gymnastics team and is a world-recognized figure in this discipline. Having been a professional dancer and director of an art company, she is also a scientist specializing in coaching sciences applied to rhythmic gymnastics.

# FOREWORD

From my earliest steps in gymnastics I have worked in the development of different aspects of this magical sport and have never deviated from this path.

The cultural aspects, the aesthetic appeal and the underlying science are among the features of gymnastics that I value. Indeed, human beings' development is based on their movement capacity. The variety of their motor skills allows them their 'freedom of movement' as well as a wide range of creativity.

Nowadays competitive gymnastics is flourishing. It is present at every major sporting event. It is a fascinating, dynamic and emotionally inspiring sport … . In one word, it is *artistic*.

The coaches, technicians and scientists working in the field are responsible for gymnastics' development. They always have to keep in mind that technique serves the artistic side of the sport.

You, the scientists who are the authors of this volume, have my compliments and my gratitude for your contribution to gymnastics.

**Prof. Bruno Grandi**
**President of the**
**International Gymnastic Federation**

# PREFACE

This book was born from a worldwide need for a gymnastics textbook to serve students, teachers, lecturers and coaches. Considerable research and applied work in gymnastics have been carried out since the 1970s by sports scientists and these have contributed to the building of a scientific basis for the sport. However, there is still a paucity of literature resources regarding the three main theoretical topics relevant to this demanding sport, namely physiology, biomechanics and psychology.

This book brings together, for the first time ever, the personal coaching experiences, the research and the theoretical reflections of key authors representing three continents: North and South America (through William Sands and John Salmela); Europe (through Patrice Holvoet, Maria Gateva and Monèm Jemni); and Africa (through Monèm Jemni). All these authors have proved themselves not only as former international coaches but also as world-leading authorities in their respective areas of expertise.

The first three parts of this book highlight physiological, biomechanical and psychological concepts as applied to gymnastics. The last part discusses how these three sciences interact with each other to contribute to a better understanding of gymnastics performance from both "learning" and "coaching" points of view. All these parts draw together not only basic definitions to introduce the main science/coaching concepts but also further cutting-edge and groundbreaking research linking theory to performance.

The physiology section highlights fundamental aspects of the field related to artistic gymnastics. It starts with an overview of the energetic systems and their applications, and is followed by the cardiovascular and respiratory responses during gymnastic exercises, the physical fitness model, physical and physiological assessments, training principles, nutrition and supplementation, growth and development issues. A final chapter is dedicated to some innovative investigations in rhythmic gymnastics.

The biomechanics section draws a full picture of the applications of the principles and techniques of physics and mechanics to the movement of the gymnasts and the apparatus. Relevant concepts and studies that contribute to a better understanding of complex human movements are included. Among them are kinetics and kinematics, angular and linear motions, and angular momentum.

The psychology section discusses, in very easy language, mental skills concepts applied in actual training and competitions. It brings together both theory and practice in sports psychology, including stress, anxiety and coping; motivation and goal setting; mental training for practising and competing; and the various stages of learning and performance.

Each part finishes with a brief conclusion summarizing the main concepts and closes with a useful questions section.

# ACKNOWLEDGEMENTS

The valuable comments, feedback and suggestions of several colleagues on the earlier drafts have made a significant improvement in this book. I am delighted to thank:

from the University of Greenwich, UK:

- Dr Jeff Pedley
- Dr Mark Goss-Sampson
- Dr Rob Willson

from the University of Cambridge, UK:

- Dr Rashid Zaman.

My thanks to Dr Peter Niemczuk as well for his effort in proof reading.

I would also like to thank the Routledge team for their help and assistance in making this project happen, especially Joshua Wells and Simon Whitmore.

# PART I

# Physiology for gymnastics

## Introduction and objectives

*Monèm Jemni*

Physiology is the science that explains how the body systems work and how these systems interact with each other to regulate their functions. The study of individual organs and their operational biochemistries is also part of human physiology.

Sporting performances have evolved throughout recent decades partly thanks to the application of different sciences to exercise. Among these sciences, exercise physiology made a major contribution to developing the understanding of 'how human systems work under different exercise conditions and regimes'. One of the sports that has witnessed significant expansion is gymnastics.

One of the particularities of gymnastics is that males and females compete in different events although there are some similarities. The male competition consists of a rotation between six events; in the following, so called, Olympic rotation order: floor exercise; pommel horse; rings; vault; parallel bars; and high bar. Females compete in four events only: vault; uneven bars; balance beam; and floor exercise. As in the modern triathlon or pentathlon, the duration, effort, intensity, power, strength, flexibility, speed of stretch, coordination and endurance as well as the energy required to perform in each of these events differ.

This chapter is divided into sections providing the most up-to-date physiological theories on male and female gymnasts. They start by highlighting the metabolic energy supplies during gymnastic exercises and then the cardiovascular and respiratory markers for training and performance, followed by the physical abilities required to perform this sport at a high level and the applied training principles. These first sections provide enough background to embark on a review of the most specific physical and physiological assessments of these athletes. Moreover, issues such as diet and nutritional supplementation and the everlasting dilemma of growth, sexual developmental and hormonal regulation are also discussed in the following sections. These discussions are supported by research from the current literature. The final chapter in Part I describes some contemporary investigations in rhythmic gymnastics.

# 1

# ENERGETICS OF GYMNASTICS

*Monèm Jemni*

Although gymnastics as a sport is increasingly publicized, attracting global media coverage for competitions such as the World Championships and the Olympics, it has not attracted as much scientific interest as mainstream sports. As early as the 1960s, authors reported that gymnasts were characterized by a low maximal aerobic power but a high level of strength (Horak, 1969; Montpetit, 1976; Saltin & Astrand, 1967; Szogy & Cherebetiu, 1971). This has been confirmed in a more recent review of the literature which has incorporated modern artistic gymnastics (Jemni *et al.*, 2001). Very few studies have explored the differences between the physiological responses in male and female gymnasts nor the energetics of the different gymnastic events. Factors which may have contributed to the paucity of the scientific database in this field include the complexity of the sport, the lack of adequate equipment and specific physiological testing protocols. During the 1970s it was suggested that male gymnasts, irrespective of their events, had an energy expenditure similar to that of running at 13 km/h (8 mph) on a treadmill (Montpetit, 1976). More recently, studies have shown significant different physiological, biomechanical and psychological requirements between running and gymnastics.

The aim of this chapter is to provide a current overview of the physiological and energetic requirements of modern gymnastics, taking into consideration each event separately. A current physiological profile of the modern artistic gymnast will be identified.

## 1.1 Aerobic metabolism

Aerobic metabolism, also known as aerobic respiration, cell respiration and oxidative metabolism, is a series of chemical processes by which individuals produce energy through the oxidation of different substrates in the presence of oxygen. The main metabolic substrates for this process include: carbohydrates (sugars), lipids (fat) and, very rarely, proteins which are used to produce the energy source of adenosine triphosphate (ATP). Other by-products from this process include water, carbon dioxide and heat. Aerobic metabolism is the main supplier of energy for endurance exercise which may last for hours. It is known that athletes who compete in long duration/distance events have a highly developed aerobic metabolism.

### 1.1.1 *Maximal oxygen uptake ($VO_2$ max)*

One of the indicators for aerobic metabolism is $VO_2$ max, or the maximal amount of oxygen that an individual can utilize during intense or maximal exercise. This factor is generally considered to be the best indicator of an athlete's cardiovascular fitness and aerobic endurance. It can be measured by a variety of methods, mainly in the laboratory environment using different types of exercise protocols and collecting the gases exhaled while progressively increasing the workload.

Although $VO_2$ max has a genetic component (Bouchard *et al.*, 1992), it can be increased through appropriate training. It is generally known that sedentary individuals have a lower $VO_2$ max compared to that of more active subjects. Meanwhile, among many factors, three have a major influence on $VO_2$ max: age, gender and altitude (Jackson *et al.*, 1995; Jackson *et al.*, 1996; McArdle *et al.*, 2005; Trappe *et al.*, 1996).

Most elite endurance athletes have $VO_2$ max values in excess of 60 ml/kg/min. Although such high values may indicate an athlete's potential for excellent aerobic endurance, other factors, such as, metabolic threshold, exercise economy and energetic cost are better predictors and correlate better with endurance performance (Wilmore & Costill, 2005).

### 1.1.2 *Gymnasts' $VO_2$ max*

The $VO_2$ max of international-level gymnasts reported over the last 40 years (Table 1.1) is around 50 ml/kg/min, although a higher value (around 60 ml/kg/min) was reported for the American elite female gymnasts (Noble, 1975). Table 1.1 reports some of these investigations.

Barantsev (1985) reported that gymnasts' $VO_2$ max decreases between adolescence and adulthood with average values decreasing from 53.2 ± 6.3 at age 12 to 47.2 ± 6.7 ml/kg/min at age 25. This regression is associated with an increase in both the volume and

**TABLE 1.1** Average $VO_2$ max and (SD) of different level gymnasts, from the 1970s up to 2006

| | | *n* | *Level* | *Age (yrs)* | *$VO_2$ max (ml/kg/min)* |
|---|---|---|---|---|---|
| Females | Sprynarova *et al.* (1969) | – | non-elite | – | 42.5 (3.7) |
| | Noble (1975) | 3 | elite | 16–22 | 61.8 (8.0) |
| | Montgomery *et al.* (1982) | 29 | non-elite | 11–13 | 50.0 (0.9) |
| | Elbaek *et al.* (1993) | 19 | elite | 20.1 (1.7) | 50.4 (2.9) |
| Males | Bergh (1980) | – | elite | – | 51.0 |
| | Barantsev (1985) | – | non-elite | 12–13 | 53.2 (6.3) |
| | | | | 14–15 | 50.9 (6.2) |
| | | | | 17–25 | 47.2 (6.7) |
| | Goswami *et al.* (1998) | 5 | non-elite | 24.2 (3.1) | 49.6 (4.9) |
| | Lechevalier *et al.* (1999) | 9 | elite | 17–21 | 53.1 (3.2) |
| | Jemni et al. (2006) | 12 | elite | 18.5 (1) | 49.5 (5.5) |
| | | | | | 33.4 (4.8)* |
| | | 9 | non-elite | 22.7 (2) | 48.6 (4.6) |
| | | | | | 34.4 (4.6)* |

*: Upper body $VO_2$ peak measured with an arm-cranking ergometer.

intensity of training for strength and power required for difficult technical skills. Other studies have reported that increase of anaerobic power via specific training programmes leads to a decrease in aerobic capacity. However, it appears that this effect is not evident before puberty. In fact, muscular fibre specification and selective fibre-type recruitment are less evident in children compared to adults (Bar-Or, 1984; Inbar & Bar-Or, 1977). It is well known that during puberty, individuals experience significant morphological and hormonal transformations associated with an increase of the maximal anaerobic power (Bedu *et al.*, 1991; Falgairette *et al.*, 1991). It is accepted by coaches that this peri-pubertal stage constitutes a crucial period for maximum potential in technical learning, strength and power gains. Taking into consideration all of the above, very clear and progressive periodization of the training seasons is necessary to avoid burnout.

Table 1.1 shows recent updated upper- and lower-body $VO_2$ max/peak values of male national and international gymnasts (Jemni *et al.*, 2006). This update was warranted by the important evolution and transformation of artistic gymnastics since the 80s. Gymnasts' upper-body $VO_2$ peak represents two-thirds of that measured on the treadmill (around 35 ml/kg/min). This value is quite high; the reason might be the fact that four out of six events in male artistic gymnastics involve exclusively the upper body (pommels, rings, parallel and high bars).

The rules of the competitions (Code of Points), as well as the compulsory elements and the free routines, are changed in every Olympic cycle, with greater emphasis given to strength and power-based skills. Consequently, both training and preparation for competitions have changed in order to match the new requirements and rules. Nowadays, gymnasts spend more time training than in the past. Training intensity and volume have increased considerably, especially at the highest levels where training often exceeds 34 hours per week (Richards *et al.*, 1999). However, the average $VO_2$ max values of both elite and non-elite gymnasts have not changed appreciably for the last five decades, remaining around 50 ml/kg/min (Jemni *et al.*, 2006) – see Table 1.1. Moreover, comparing elite and non-elite $VO_2$ max, has not disclosed any significant statistical difference, even with the large disparity in training volumes between the two levels (Jemni *et al.*, 2006).

Conventional wisdom would indicate that the elite group should have a greater aerobic capacity, due to higher practice volumes, although, unlike in running, swimming, and cycling, $VO_2$ max in gymnasts is not directly related to performance parameters. Gymnasts' $VO_2$ max values have been reported to be significantly lower than those of athletes participating in short, intense activities, such as high-level male sprinters (Barantsev, 1985; Bergh, 1980; Goswami & Gupta, 1998; Marcinik *et al.*, 1991), and some reports suggested that they are comparable to values for the sedentary population (Crielaard & Pirnay, 1981; Wilmore & Costill, 1999).

Finally, coaches' opinions differ concerning gymnasts' $VO_2$ max. Many coaches are still encouraging gymnasts to develop their $VO_2$ max by planning jogging sessions and/or long activities such as cycling on gym ergometers. They justify this training with the claim that 'having an important oxidative foundation enhances recovery during high-intensity sessions'. Other coaches still believe that long 'endurance-type sessions' help weight control, especially in female gymnasts (Sands, McNeal, & Jemni, 2000). There is evidence that aerobic endurance training may interfere with power, the main physical quality of a gymnast. This section clearly demonstrates that gymnasts' $VO_2$ max have not changed appreciably for the last five decades (remaining around 50 ml/kg/min) while their performance has increased significantly. The findings provide strong evidence that it is unnecessary to enhance gymnasts' $VO_2$ max.

### 1.1.3 Gymnasts' metabolic thresholds

Several metabolic thresholds commonly assessed during maximal incremental exercise tests are considered to be better predictors of endurance performance than the $VO_2$ max (Wilmore & Costill, 2005). These include the anaerobic threshold (AT), lactate threshold (LT), lactate turn point (LTP), onset of blood lactate accumulation (OBLA) and the ventilatory threshold (VT). It is to be expected that high-level endurance athletes would have delayed metabolic thresholds as the consequence of an improved oxidative system. Few studies have investigated these metabolic thresholds in gymnasts. Table 1.2 shows the LT assessed in elite and sub-elite male artistic gymnasts during a maximal incremental test performed on the treadmill (Jemni *et al.*, 2006). Both categories of gymnasts achieved their LT at a very high percentage of their $VO_2$ max (mean 79%) and at a very similar heart rate (mean 169 bpm, or beats per minute).

A low $VO_2$ max, as described in the previous section in association with a high metabolic threshold, is not a common feature of a conventional physiological profile. Analysis of this feature will be described progressively throughout this chapter.

Increasing training volume in gymnastics would be expected to result in physiological adaptations that reflect an improvement in the removal of blood lactate through increased buffering capacity, increased uptake through neighbouring slow-twitch fibres or increased removal through the Cori cycle. Alternatively, there might be less production, as a greater training volume might enhance the capacity to sustain a higher intensity by relying on limited anaerobic energy sources. This is suggested by the low blood lactate values measured in gymnasts at the end of the maximal exercise test and by the high percentage of $VO_2$ achieved at the lactate turn point (around 10 mmol/l and 79% respectively) (Jemni *et al.*, 2006). Interestingly, the lactate turn point of gymnasts is comparable to that achieved by elite endurance athletes (Wilmore & Costill, 1999, p. 137). Long-distance runners and cyclists reach their lactate turn point at similar levels (77% of $VO_2$ max in a group of endurance cyclists measured by Denadai *et al.* (2004), for example). Several authors have noted a relationship between increased strength and increased anaerobic power and measures of endurance as a result of strength training (Hickson *et al.*, 1988; Marcinik *et al.*, 1991). It has been demonstrated that lactate threshold can be delayed markedly through resistance training (Marcinik *et al.*, 1991). Jemni *et al.* (2006) have proved that the amount of strength and conditioning performed during gymnastic training not only increases strength and power but also enhances fatigue tolerance.

**TABLE 1.2** Average male gymnasts' lactic thresholds and (SD) assessed during maximal incremental test performed on the treadmill (Jemni *et al.*, 2006)

| | *HR (bpm)* | *$VO_2$ max (ml/kg/min)* | *% $VO_2$ max* |
|---|---|---|---|
| Elite level (n = 12, 18.5 yrs) | 169.8 (10.1) | 44.5 (5.4) | 82.1 (6.5) |
| Non-elite (n = 9, 22.7 yrs) | 169.3 (4.0) | 37.0 (5.1)$^S$ | 76.3 (9.9) |

S ($p < .05$): significant difference.

## 1.2 Energy cost of gymnastics exercises

The energy cost of gymnastics was first measured in the early 50s in the old Eastern bloc (Blochin, 1965; Krestovnikov, 1951). Different procedures have been used to collect exhaled gas, the main ones using Douglas bags. Seliger *et al.* (1970) have supposed that such heavy equipment effects might have underestimated the energetic cost by 10%. However, technological evolution in lab equipment has allowed higher precision in measurements. Investigation methods and techniques have differed between the studies, but in all cases it is agreed that energy cost differs between events (for males and females) (Table 1.3).

Hoeger & Fisher (1981) measured the energy cost in male gymnasts performing their six compulsory routines. Gymnasts performed their six routines while equipped with a mouthpiece linked to a Douglas bag. At the end of each routine, gymnasts had to hold their breath for a few seconds until the investigators attached another bag to collect the expired air during recovery. The bags were analysed using a pneumograph MTG and a Parkinson Cowan CD 4 gas meter (Hoeger & Fisher, 1981). Results showed that the most costly event was the floor exercise (37 kcal), followed in decreasing order by the pommels horse, still rings, horizontal bar parallel bars and finally the vault. Following this, Rodríguez *et al.* (1999) measured the excess post-exercise oxygen consumption (EPOC) during the first 30 seconds following female gymnastic routines, using online gas analysis, which allowed them to estimate the real oxygen consumption. Their results were similar to those for the male routines: the floor exercise induced the highest value (40.8 ± 4.0 ml/kg/min) and the vault exercise induced the least oxygen consumption (34.3 ± 7.7 ml/kg/min). The findings of these two investigations are in agreement with the blood lactate values of male and female routines measured in other studies (Jemni *et al.*, 2000; Montgomery & Beaudin, 1982; Montpetit, 1976) (see Section 1.3.2).

Some studies have indirectly estimated the energy cost of gymnastic exercises by using the regression relation between heart rate and oxygen consumption measured during a maximal graded test and heart rate values measured during gymnastics (Noble, 1975; Montpetit, 1976).

**TABLE 1.3** Average energy cost and (SD) of male and female gymnastics events

| *Females* | *Vault* | *Uneven bars* | *Balance beam* | *Floor* |
|---|---|---|---|---|
| Seliger *et al.* (1970) ml/kg/min of $O_2$ | 16.9 (3.5) | 16.5 (3.6) | 15.1 (4.9) | – |
| Rodríguez *et al.* (1998) ml/kg/min of $O_2$ | 34.3 (7.7) | 36.6 (4.6) | 31.3 (6.1) | 40.8 (4.0) |

| *Males* | *Floor* | *Pommels* | *Rings* | *Vault* | *Parallel bars* | *Horizontal bar* |
|---|---|---|---|---|---|---|
| Ogawa *et al.* (1956) kcal/min | – | 7–11 | 11–16 | 5–10 | 8–10 | 12–15 |
| Seliger *et al.* (1970) ml/kg/min of $O_2$ | 20.5 (6.3) | 16.4 (2.5) | 17.3 (2.7) | – | 17.1 (3.6) | 18.5 (3.4) |
| Hoeger *et al.* (1981) kcal/routine | 37.01 | 36.6 | 32.5 | 25.8 | 32.3 | 32.5 |
| Sward (1985) cal/kg/min | 0.13 | 0.16 | 0.08 | 0.13 | 0.09 | 0.14 |

**TABLE 1.4** Averages and standard deviations of the male and female routines duration measured in different international competitions (in seconds) (Jemni *et al.*, 2000)

| | *Males* | *Females* |
|---|---|---|
| Floor exercise | 60.9 (3.5) | 82.9 (3.2) |
| Pommel horse | 30.5 (4.5) | – |
| Rings | 40.7 (5.1) | – |
| Vault | 5.2 (0.5) | 4.8 (0.9) |
| Parallel bars | 31.2 (6.2) | – |
| High bars | 36.5 (6.6) | – |
| Uneven bars | – | 46.5 (3.5) |
| Balance beam | – | 81.8 (4.5) |

The findings suggest that the contribution from aerobic metabolism was 20% and from anaerobic metabolism 80%. It was further suggested that gymnasts would use only 35% of their aerobic capacity as measured during the $VO_2$ max test to perform gymnastic routines. However, this relation is valid only when measurements are taken in a steady state (heart rate and oxygen uptake). It is almost impossible to achieve such a steady state while performing gymnastic routines, owing not only to the short and different durations of the routines (Table 1.4) but also to the different intensity/pace and rhythm changes. Chapter 2.5.2. shows examples of heart rate fluctuations during male routines. The steady state is difficult to achieve because of the variety of muscle contractions that gymnasts use while performing. In addition, gymnasts often perform some elements while maintaining an apnoea for a few seconds. It has been shown that apnoea has an effect on the cardiovascular system (Shaghlil, 1978).

It is possible to achieve a steady state quite quickly while running or cycling at a steady pace (around three minutes), whereas it is impossible in gymnastics because of all the above factors. Therefore, extrapolating gymnastics energy expenditure from the regression relation between heart rate and oxygen consumption measured during a maximal graded test to heart rate values measured during gymnastics, as some authors have previously done, is totally invalid.

Figure 1.1 shows the energy supply continuum. Whether activity is aerobic or anaerobic, the three processes which are responsible for ATP production – the anaerobic ATP-phosphocreatine (ATP-PC) system, the anaerobic glycolysis system and the aerobic oxidation system – work together rather than in sequence. It is well known that at a certain time during the exercise, one of these three processes would be the main source of ATP, depending on the duration and the intensity. Short and very intense burst like exercises rely predominantly on the anaerobic ATP-PC system, which provides an immediate but very limited source of ATP for explosive contractions and power development, as required for the vault. Restoration of the ATP-PC system is rapid during the following post-exercise rest period, and requires oxygen supply and very great reduction of exercise intensity.

If the exercise is intended to be continued after a very powerful start lasting a few seconds (as is the case for a 400-m sprinter), then it is generally expected that the intensity of the exercise would drop slightly. In the meantime the energy supply would progressively rely on anaerobic glycolysis, which can be fully functioning around 20 seconds. At this stage, the high intensity of the exercise can only be maintained for a short period because of the lack of oxygen and the considerable production of lactic acid via anaerobic glycolysis. This hypoxic

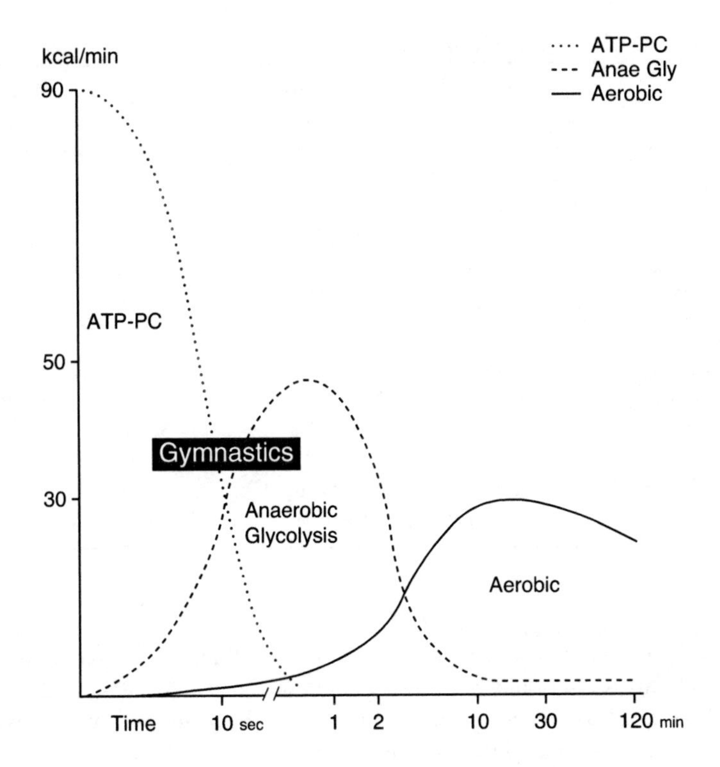


**FIGURE 1.1** The energy continuum. (Adapted from McArdle, Katch, & Katch, 2005).

environment restricts the efficacy of the muscle contractions and therefore muscular fatigue and power output declines are the main outcomes. However, it is important to understand that muscle fatigue is not the direct result of lactate accumulation at this stage. Powerful athletes are able to push the boundaries of the high acidity with little drop in the efficacy of their exercise. Anaerobic and resistance training would enhance this specific performance. The anaerobic system is considered as the main energy supplier of most of gymnastics' competitive events. Blood lactate measurements might reflect the contribution of this metabolism, as shown in Section 1.3.2. There is no doubt that the intensity of the competitive events is very high. This intensity is pushed to the extreme in high-level gymnasts. Nowadays, it is common to see gymnasts performing double front or back somersaults including twists on the vault or floor exercises, or as dismounts from several types of apparatus. In addition, the longest events are the floor exercises and the balance beam (90 seconds maximum for female floor exercises and balance beam; 70 seconds maximum for male floor exercises). Pommels, rings, parallel bars, uneven bars and horizontal bar routines last approximately 35 seconds, whereas a vault lasts only 6 seconds on average.

As shown in Figure 1.1, if the physical exercise has to be continued beyond anaerobic glycolysis, the body would mainly rely on the oxidative pathway system for the energy supply. A corresponding drop in the intensity would also occur. This system is the main energy supplier for endurance sports lasting for longer periods (Section 1.1). However, in gymnastics all the events are short and require quick explosive movements. Although the oxidative system may contribute slightly in several events, their short durations would not allow a full energy supply. Indeed the oxidative system requires a few minutes to be fully functional and to act as the full provider of the energy during aerobic activities.

**TABLE 1.5** Estimation of energy supply in each gymnastic event

| | *ATP-PCr* | *Anaerobic glycolysis* | *Oxidative* | *Blood lactate (mmol/l)* |
|---|---|---|---|---|
| *Females* | | | | |
| Vault (6 sec) | 100% | 5–10% | 1–2% | 2.5* |
| Uneven bars (45 sec) | 100% | 80–90% | 3–5% | 7.4* |
| Balance beam (90 sec) | 90% | 50–60% | 20–30% | 4.3* |
| Floor exercises (90 sec) | 100% | 80–90% | 20–30% | 7.0* |
| *Males* | | | | |
| Floor exercises (70 sec) | 100% | 60–70% | 20–30% | 6.2** |
| Pommel horse (35 sec) | 100% | 80–90% | 3–8% | 5.8** |
| Rings (35 sec) | 100% | 80–90% | 3–8% | 5.8** |
| Vault (6 sec) | 100% | 5–10% | 1–2% | 3.8** |
| Parallel bars (35 sec) | 100% | 70–80% | 5–10% | 4.0** |
| High bars (35 sec) | 100% | 70–80% | 3–8% | 5.0** |

*: (Rodríguez *et al.*, 1999); **: (Jemni *et al.*, 2000).

Unfortunately, no one has so far attempted to perform biopsies on gymnasts, mainly for ethical reasons. This is essentially the most accurate technique to assess energy production and supply. Table 1.5 shows energy estimation in each of the gymnastic events based on their respective duration and on blood lactate assessment.

## 1.3 Anaerobic metabolism

Anaerobic metabolism is generally the energy provider for short bursts of high-intensity exercise lasting a few minutes. The energy is produced through a series of chemical reactions called glycolysis. It occurs in the absence of oxygen and involves the breakdown of carbohydrates (glucose, glycogen) as a fuel source. The anaerobic metabolism is limited in time by the accumulation of protons ($H^+$) and the increase of acidity in the muscle cells within a few minutes, because the rate of production of lactic acid exceeds its removal rate. Lactic acid is in fact a natural by-product of anaerobic glycolysis. The acidic environment impairs the homeostasis balance by increasing the concentration of the H+ in the blood. The body's buffering systems' activities (chemical, respiratory, urinary) could also be impaired and lead to a reduction in the muscle tension. This leads to a progressive reduction in physical performance and to exhaustion. The impairment of the muscle contractions might affect technical performance, especially in artistic sports such as gymnastics. Gymnasts are indeed marked on 'how well they perform', not on how many skills they are able to present. Any imperfection is penalized – even a single step at a landing of an acrobatic element. It is therefore fundamental that gymnasts are able to perform under a high level of 'metabolic stress', given the short duration of their routines and the high intensity levels of their exercises.

Many authors have reported that modern artistic gymnastics requires greater strength and power because of the ever increasing technical difficulty required through revision of the International Gymnastics Federation (FIG) Code of Points (FIG, 2009; Brooks, 2003; French *et al.*, 2004). FIG, the international governing body, reviews and updates the code of judging every four years. Hence, performance demands on gymnasts are continually changing to meet the new Code requirements. In the 70s, the Code of Points showed only three levels of difficulty: A, B and C. In 2009, the Code of Points showed not only an increase in the number of technical skills but also seven levels of difficulty: A to G. Routines which include E, F and G skills have higher start values than routines constructed only with B, C and D skills. Gymnasts are encouraged from an early age to learn more difficult skills to ensure a higher start value that leads to a higher score in order to reach the highest levels of competition. This suggests that modern artistic gymnasts have to develop their anaerobic metabolism at an early age in order to perform at the highest technical levels. Nowadays, it is very common to see two-year-old children in a gymnasium initiation programme, because it typically takes about 10 years for a gymnast to reach the elite level.

This chapter highlights several components of gymnasts' anaerobic metabolism.

### *1.3.1 Power output of gymnasts*

The power of the anaerobic metabolism can be assessed through a series of biochemical analyses following a muscle biopsy. This technique is indeed the best approach to evaluate the physiological mechanisms that underpin a physical performance. Unfortunately, this technique is not only painful but also costly and requires highly qualified technicians and a medical facility. Indirect, non-invasive techniques can give an idea of the power of the anaerobic metabolism. Ergometry results have been strongly correlated with the outcomes of the invasive techniques and therefore are widely accepted as means to estimate anaerobic metabolism.

Currently, there is no specific ergometer to assess gymnasts. Sport scientists commonly apply standardized ergometry tests in the laboratory or in the gym. The Force–Velocity test (Vandewalle *et al.*, 1987) and the Wingate test (Bar-Or, 1987) are considered gold standards to estimate mechanical power output and are the ones used most frequently. These tests might be performed by the upper and/or the lower body limbs. The Force–Velocity test consists of measuring peak cycling velocity during short maximal sprints (about six seconds) at different resistances. The Wingate test consists of measuring not only the peak power generated during the first 10 seconds of a full out 30-second cycling exercise but also the mean power generated during the same period. Regrettably, gymnasts are not used to cycling or arm cranking and therefore they are somewhat disadvantaged compared to other athletes. However, performing such recognized standard tests allows for comparisons with other athletes.

Tables 1.6 and 1.7 show the results of Force–Velocity and Wingate tests respectively, performed by elite and non-elite gymnasts. Note there is a statistically significant difference

**TABLE 1.6** Average upper- and lower-body peak power output and (SD) measured in male gymnasts during Force-Velocity test (Jemni *et al.*, 2001; Jemni *et al.*, 2006)

| | | *[Peak power]$_{6sec}$ (W)* | *[Peak power]$_{6sec}$ (W/kg)* |
|---|---|---|---|
| Elite level (*n* = 12, 18.5 yrs) | Upper body | 688.3 (87.7) | 10.6 (0.9) |
| | Lower body | 1028.0 (111.5)$^S$ | 15.9 (1.3)$^S$ |
| Non-elite (*n* = 9, 22.7 yrs) | Upper body | 652.4 (79.9) | 9.8 (1.1) |
| | Lower body | 980.7 (266.4)$^S$ | 15.1 (4.3)$^S$ |

S ($p < .05$): significant difference between upper and lower body results but no difference between the levels.

between the upper- and lower-body performances within the same level, whereas there are no statistically significant differences between the levels of practice. Maximal power output relative to body mass developed by the upper body of the gymnasts represents two-thirds of that developed by the lower body in both levels.

These high peak-power values (~15 W/kg in the Force–Velocity, ~13.5 W/kg in the Wingate for males and ~10.5 W/kg in the Wingate for the females) place the gymnasts near the top levels of power athletes. For example, upper- and lower-body powers of the male gymnasts are higher than those measured in elite wrestlers (7.8 ± 1 W/kg and 10.9 ± 1.2 W/kg, respectively for upper and lower body) (Horswill *et al.*, 1992).

When comparing the force–velocity results with those from similar investigations using the same protocol with other sport groups, male gymnasts have similar upper-body values to swimmers – ~10.5 W/kg according to Vandewalle *et al.*, 1989; similar lower-body values to volley-ball players – ~15.8 W/kg (Driss *et al.*, 1998); and just below the lower-body values scored by sprinters – ~17.0 W/kg (Garnier *et al.*, 1995).

It is evident that 'peak power' is a key component in the gymnast's physiological profile. In fact, the contribution of strength and power to gymnastic performance has increased during the last four decades (Jemni *et al.*, 2001). The total effort time of a gymnast during a three-hour competition is only between 12 and 15 minutes (warm-up included) (Jemni *et al.*, 2000). During these minutes of effort time, gymnasts must perform a limited number of skills in a limited time (at each event). To achieve powerful skills, speed and strength are indeed key components, especially in modern artistic gymnastics. The specificity of gymnastics training has certainly had an effect on shaping the energetic requirements of the practitioners. The literature provides evidence that specific training has an influence on physical aptitudes, on fibre-type characteristics and, indirectly, on aerobic and/or anaerobic metabolism (Jansson *et al.*, 1978). Clearly, repeating powerful gymnastics skills and carrying and/or supporting the body weight on the apparatus would enhance the above qualities. In addition, Table 1.7 shows high blood lactate values measured after Wingate tests performed with the upper and lower bodies in male and female gymnasts (around 10.5 mmol/l). These high values are indirectly symptomatic of an established anaerobic metabolism. More detail about this metabolism is highlighted in the following section through investigations in blood lactate during gymnastic routines.

### *1.3.2 Blood lactate measurement during gymnastic exercises*

Blood lactate (BL) analysis allows an indirect estimation of the anaerobic glycolysis contribution. Sadly, there have been few attempts to measure BL during gymnastics routines, in particular in male gymnasts. In the 1970s some authors supposed that lactate production was negligible (Montpetit, 1976), suggesting that anaerobic glycolysis is not the main supplier of energy and that the main part of energy production is assured by the ATP-PC system. Beaudin (1978) reported an average BL of 2.8 mmol/l following the four female routines, with higher values measured after the floor exercises and the uneven bars (~4.4 mmol/l). These low values might be attributed to the low level of the gymnasts, who were unable to perform high-difficulty exercises. Higher BL values were reported in the 80s and in the 90s for low and higher ability levels, with Montgomery & Beaudin (1982) reporting an average of 4.0 mmol/l and Rodríguez *et al.* (1999) an average of 5.3 mmol/l (Table 1.8). Also, Goswami & Gupta (1998) and Le Chevalier *et al.* (1999) found similar averages in male gymnasts despite the difference in their research methods (6.2 ± 0.7 and 6.2 ± 1.6 mmol/l respectively, on five types of apparatus, the vault being excluded).

**TABLE 1.7** Average upper- and lower-body power outputs and (SD) measured in male and female gymnasts during Wingate test (Jemni, 2001; Jemni *et al.*, 2006)

| | | MPO (W/kg) | PPO (W/kg) | BL max (mmol/l) |
|---|---|---|---|---|
| *Females* | | *Lower body* | | |
| Heller *et al.* (1998) | Elite level (n = 6, ~15.5 yrs) | 8.6 (0.1) | 10.4 (0.4) | 11.6 (1.7) |
| Sands *et al.* (1987) | (n = 25, ~14 yrs) | 7.1 (1.3) | 7.9 (2.0) | – |
| | | *Upper body* | | |
| Sands *et al.* (1987) | (n = 25, ~14 yrs) | 3.1 (0.7) | 3.6 (1.0) | – |
| *Males* | | *Lower body* | | |
| Jemni *et al.* (2006) | Elite level (n = 12, 18.5 yrs) | 9.7 (1.00)[s] | 13.5 (1.34) | 11.7 (2.03) |
| | Non-elite (n = 9, 22.7 yrs) | 10.1 (1.3)[s] | 14.1 (3.0) | 11.0 (3.1) |
| Savchin *et al.* (2003) | Mixed (n = 24, 11.6 yrs) | – | 9.6 (1.9) | 12.1 (0.9) |
| Heller *et al.* (1998) | Elite level (n = 5, >18 yrs) | 10.7 | 13.2 (1.0) | 11.2 (1.5) |
| | | *Upper body* | | |
| Jemni *et al.* (2006) | Elite level (n = 12, 18.5 yrs) | 7.1 (0.5)[s] | 9.6 (0.6) | 12.2 (1.5) |
| | Non-elite (n = 9, 22.7 yrs) | 6.6 (0.6)[s] | 9.2 (1.1) | 10.4 (0.7) |

MPO: Mean Power Output; PPO: Peak Power Output; BL: Blood Lactate post Wingate test; *n*: number;
S ($p < .05$): significant difference between upper and lower body results but no difference between the levels.

Figure 1.2 shows the maximum and minimum blood lactate values (BL) assessed during male gymnastics competition. The competition was composed of the six Olympic events and began with the floor exercise, followed by pommel horse, rings, vaulting, parallel bars and horizontal bar. The routines were separated from each other by 10 minutes of recovery. The programmes performed were free but technical composition did not vary between gymnasts.

The average maximum BL over all six events was 4.8 ± 1.1 mmol/l. This average would be even higher if the vault values did not count (5.1 ± 0.9 mmol/l).

**TABLE 1.8** Blood lactate for each event in male and female gymnasts in mmol/l

*Females*

| | Montgomery *et al.* (1982) (n = 29) non-elite | Bunc *et al.* (1994) (n = 7) elite | Rodríguez *et al.* (1999) (n = 8) mixed |
|---|---|---|---|
| Vault | 3.1 | 2.4 | 2.5 |
| Uneven bars | 2.2 | 9.5 | 7.4 |
| Balance beam | 3.0 | 10.2 | 4.3 |
| Floor exercises | 8.5 | – | 7.0 |

*Males*

| | Goswami *et al.* (1998) (n = 5) non-elite | Jemni *et al.* (2000) (n = 7) non-elite | Groussard *et al.* (2002) (n = 5) mixed | Jemni *et al.* (2003) (n = 12) mixed |
|---|---|---|---|---|
| Floor exercises | 7.11 | 6.2 | 11 | 6.3 |
| Pommel horse | 5.18 | 5.8 | 6.5 | 6.0 |
| Rings | 6.77 | 5.8 | 6.6 | 6.7 |
| Vault | – | 3.8 | 5.0 | 3.8 |
| Parallel bars | 6.23 | 4.0 | 5.8 | 5.1 |
| High bars | 5.97 | 5.0 | 6.0 | 5.2 |

As shown in Figure 1.2, BL varied from one apparatus to another. The highest value was observed for the floor exercise (6 to 11 mmol/l); this was significantly higher than those of vault, parallel bars and the horizontal bar. Blood lactate measured for the vault (3 to 4 mmol/1) was significantly less than those of the other exercises. These two findings confirm similar investigations performed on female gymnasts (Rodríguez *et al.*, 1999). Figure 1.2 also indicates

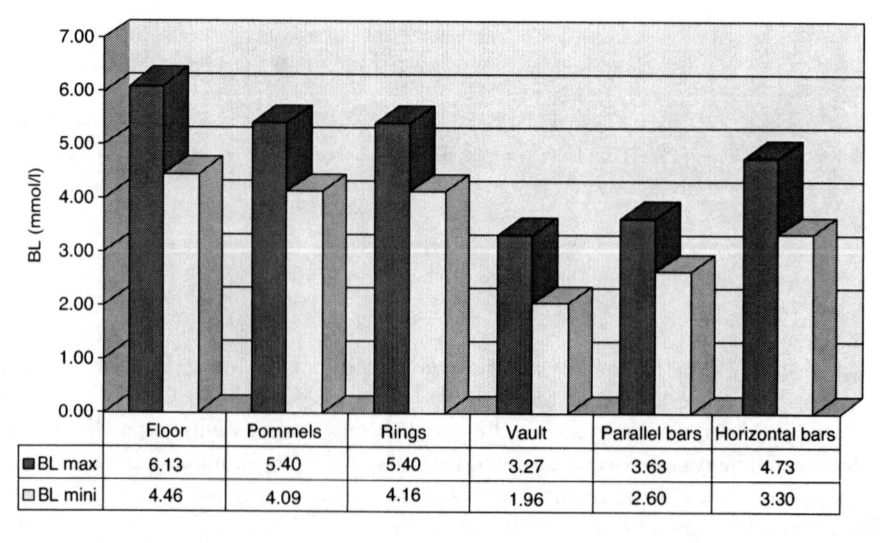


**FIGURE 1.2** Blood lactate during males' gymnastics competition in mmol/l (Jemni *et al.*, 2000).

that even BL decreased after the 10-minute recovery period; the values were always higher than the expected rest values (less than 2 mmol/l).

This average could not be considered as negligible as it is above the Onset of Blood Lactate Accumulation of 4 mmol/l (OBLA), indicating the moment when the rate of lactate production becomes more important than its removal. This also indirectly indicates the moment when anaerobic metabolism becomes the main energy supplier.

Coaches pursue varying training objectives based on the particular period of training. Gymnasts, much like sprinters, perform in both aerobic and anaerobic conditions (Jemni *et al.*, 2000; Sands, 1998). During intense sessions, gymnasts are asked to perform routines while fatigued. They are often asked to find the best compromise between technical effectiveness, safety, and high intensity effort. During the competitive phase of the season, gymnasts usually repeat their six events several times per practice session (Arkaev & Suchilin, 2004). Figure 1.3 is a typical example of different practice sessions. Blood lactate was measured during 1) 'an easy session', which was mainly rehearsing skills on three types of apparatus (floor, pommel horse and rings); 2) a competition where the gymnasts performed their six routines; 3) double competition, where the gymnasts performed their six routines twice (also called 'back-to-back'). This figure indicates that anaerobic metabolism becomes a major contributor when the gymnasts have to repeat their routines more than once. The average BL in the double competition was 5.8 ± 1.9 mmol/l (6.3 ± 1.7 mmol/l for five types of apparatus, excluding the vault). This value might increase considerably if the gymnasts are requested to repeat their routines more than twice. These types of practices are similar to interval training due to the intermittent and intense activities that are involved. Therefore one of the questions that should be addressed is: how should gymnasts recover in these types of practice? (See Chapter 4.6). Figure 1.3 shows also that BL was very low during the 'skill learning session' (2.0 ± 0.7 mmol/l).

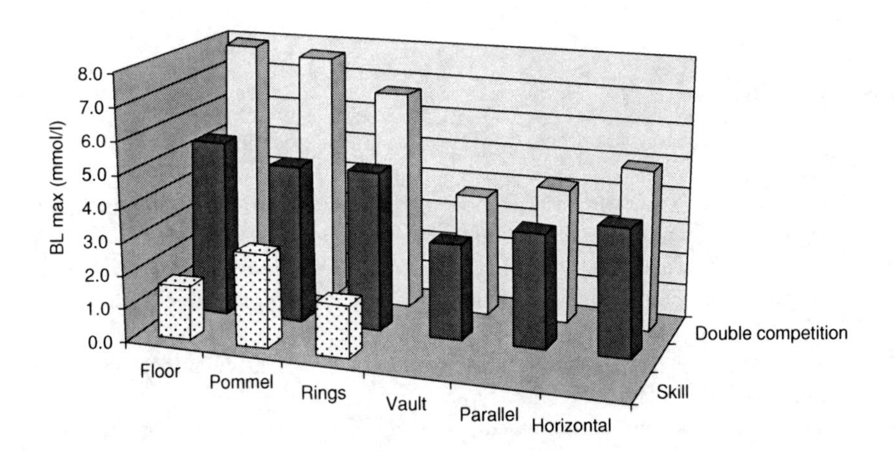


| | Floor | Pommel | Rings | Vault | Parallel | Horizontal |
|---|---|---|---|---|---|---|
| ▩ Skill | 1.6 | 2.8 | 1.6 | | | |
| ■ Competition | 5.4 | 4.8 | 4.9 | 2.9 | 3.5 | 3.9 |
| □ Double competition | 7.9 | 7.7 | 6.7 | 3.7 | 4.1 | 4.9 |

**FIGURE 1.3** Blood lactate in mmol/l measured in three different males' gymnastic sessions: Skills learning, Competition, Double competition (Adapted from Lechevalier *et al.*, 1999).

In addition, average BL was similar to the one shown in Figure 1.2 (4.2 ± 1.0 mmol/l for six events or 4.5 ± 0.8 mmol/l in five events).

Interestingly, by rehearsing the same skills/routines, the gymnast becomes increasingly 'economical'. The gymnast progressively finds the most efficient movements and indirectly the suitable contractions allowing him/her to perform the exercises without undue fatigue. Gymnastics practice is indeed based on 'repetition'. High-level gymnasts often repeat more than 1700 elements per micro-cycle (7 to 8 per day), not counting the strength and conditioning exercises (Arkaev & Suchilin, 2004). They have to master each element separately before including it with a mini-routine or a combination of elements. Learning these skills separately is indeed not only the way to refine the technique, but also another way to find the best haemodynamics, allowing minimum energy expenditure. One would expect to see the energy expenditure of a gymnast reduce between the start and the end of a learning period. It is also suggested that lactate production would also be reduced. In addition, a high-level gymnast would produce less lactate compared to a lower-level gymnast if they are asked to perform the same task. The high-level gymnast would use less force and would rely on his most 'economical' muscle fibres and therefore would produce less lactate. The lower-level gymnast would use a higher percentage of his strength capacity, he would rely more on type $II_B$ fibres and thus would produce more lactate. Moreover, one of the specific effects of training is the enhancement of 'lactate tolerance'. Sands (2003) suggested that because of the risk of injuries associated with practice, maximum levels of lactate production are not likely to be achieved due to the increasing acidity, contractile failure and indirect fatigue. The author concluded that the maximum 'safe' level of lactate for competent skill performance is still unknown. However, lactate tends to mirror the effects of fatigue and can be used to describe the magnitude of glycolytic involvement.

## 1.4 Conclusion

The following conclusions are drawn from the previous two sections:

- There is evidence that gymnasts' aerobic metabolism is classified among the lowest compared to other athletes and is similar to that of non-active persons.
- Gymnasts' $VO_2$ max has not increased for the last five decades in spite of an increased training volume and intensity in conjunction with 'tougher' rules (~50 ml/kg/min).
- There is evidence that gymnastic practice is not a sufficiently significant stimulus to enhance the oxidative system.
- Gymnasts demonstrate an increased maximal power output as measured by standardized tests such as force–velocity and Wingate tests (up to 16 W/kg in males and up to 14 W/kg in females). This classifies them among the most powerful athletes.
- Gymnasts' metabolic thresholds are achieved at an early stage during maximal incremental tests (~80% of their $VO_2$ max). This delay is the result of the important volume of strength and conditioning.
- There is a paucity of literature resources regarding energy cost of male and female gymnastic routines. Meanwhile, there is evidence showing the increasing contribution of the anaerobic metabolism (blood lactate up to 11 mmol/l).

# 2

# CARDIOVASCULAR AND RESPIRATORY SYSTEMS OF GYMNASTS

*Monèm Jemni*

## 2.1 Respiratory and ventilation system

There is an evident paucity of literature regarding the respiratory system in male and female gymnasts. This lack of information is mainly due to the difficulties of operating medical instruments while performing gymnastics. It is indeed very tricky, even with advanced technology, to collect the gas exchange during aerial catch and release skills. This is the case, for example, on the horizontal and parallel bars. Investigators have tried to use portable telemetric gas analysers; unfortunately, the weight of the equipment obliged the gymnasts to make some adjustments to the skills. Aerial figures, for example, were harder to perform. The gymnasts were not only afraid of missing their landings/dismounts and thereby damaging the equipment, but they were also prevented from performing certain elements where they had to go on their backs and/or their fronts. In addition, the face mask and the turbine of the system prevent head flexion and tucked or piked positions on all apparatus (as in the case of somersaults).

The only two studies reported in the literature within this field were carried out in the late 70s. The authors gave details mainly of the ventilatory system of the gymnasts at rest: Shaghlil (1978) showed that the breathing frequency in gymnasts at rest was lower than that measured in a group of sedentary subjects of a similar age (12 to 14 versus 16 to 18 breaths/min respectively). Gymnasts' ventilation varies, particularly in certain situations such as the handstand, when respiration becomes more difficult due to the viscera putting more weight on the diaphragm. This would increase the intra-thorax pressure, slow down the ventilation and cause congestion of the neck and the face. However, this is not due to any modifications of the characteristics or the functioning of the respiratory muscles.

Barlett *et al.* (1984) have assessed the expiratory reserve volume (ERV), expressed as a percentage of vital capacity (VC), in a group of sub-elite female gymnasts and compared it to a group of runners of a similar age (29.7 ± 7.1% vs 43.1 ± 6.4%, respectively). This significant lower ERV was associated with a significant larger upper-body composition in the female gymnasts. The authors suggested the greater upper-body mass in female gymnasts (the extra muscle on the chest) impinges upon the thorax to reduce its resting end-expiratory dimensions and, hence, the ERV of the lungs. Nevertheless, this decrease in lung efficiency is not a limiting factor in gymnasts' performances because they do not use their full lung capacity as in endurance sports. *(Note, information about maximal oxygen uptake in gymnasts is in Section 1.1.)*

## 2.2 Cardiovascular adaptation to gymnastic exercises

Practising gymnastics, like any other regular physical activity, will induce some cardiovascular adaptations; among them, a normal hypertrophy of the myocardium, a decreased resting heart rate and an increased systolic ejection volume. However, very few studies have investigated the cardiovascular adaptations to gymnastics training and, therefore, extensive conclusions cannot be drawn at this stage.

According to Potiron-Josse & Bourdon (1989), regular exercise increases the cavity dimensions of the heart by 30% in active adults compared to sedentary people. Nonetheless, Roskamm (1980) has compared the volume of the heart relative to body mass of different athletes from the German teams with an age-matched group of inactive persons. He concluded that weightlifters have the lowest volume followed by inactive individuals and gymnasts (10.8 ml/kg and 11.7 ml/kg respectively). Similar findings have been confirmed by a later study conducted by Obert *et al.* (1997); their comparisons of pre-pubertal gymnasts with similarly aged non-gymnast children did not show any significant differences for the following variables: myocardial mass, systolic diameter, systolic and diastolic ejection fractions, cardiac output, heart rate, and systolic volume.

According to Shaghlil (1978), the blood pressure of gymnasts does not differ from that of sedentary people of the same age. However, a slight increase might occur at the approach of competitions. It is indeed during this critical period when the physical, physiological and psychological profile of the gymnast would take its 'competitive shape'. The increasing competitive anxiety might be one of the reasons behind the slight increase in blood pressure. The same author confirmed that during handstands, acrobatic elements and full swings at the horizontal bar, some modifications of local blood flow occur due to the centrifugal and/or centripetal force. However, the vascular system conserves a fair distribution of the blood volume. In addition, local blood flow is quickly restored to normal after exercise.

## 2.3 Cardiac response during gymnastic exercises

Among the most recent investigations of cardiac response, Montpetit & Matte (1969) have shown a significant decrease of the heart rate (HR) when holding a handstand for 30 seconds. HR drops from 120 bpm and stabilizes around 95 bpm after 5 seconds. It increases slightly after the cessation of the exercise to finally drop back to the original level after 10 seconds. This decrease is explained by an increased stroke volume following a sudden increase in venous return and the reverse occurs when returning to the standing position.

Seliger *et al.* (1970) and Faria & Pillips (1970) were also among the first who studied cardiac responses in varieties of gymnastic routines. Thanks to the electrocardiogram, Seliger *et al.* (1970) were able to assess the cardiac stress in male and female gymnasts. Heart rates reached 148 bpm on the balance beam, 135 on the uneven bars and 133 in the vault. Meanwhile, it was slightly higher in males; it ranged between 139 bpm on the parallel bars and 151 bpm on the floor exercises. These values seem to be quite low and can be explained by the nature of the 'easy' exercises of the 1970s (Table 2.1). During the same period, Noble (1975) was able to measure the cardiac response among female gymnasts using electrocardiographs emitting signals every 5 seconds. The values found for the floor exercises, balance beam and uneven bars are higher than those found by Seliger. They lie between 152 and 189 bpm. The average values for these events were respectively 169 ± 6, 159 ± 6 and 167 ± 2 bpm.

In 1976, Montpetit found a fairly high cardiac response among male gymnasts performing simple exercises. A telemetric electrocardiograph was used to record the HRs. The values found ranged between 130 and 170 bpm. These are higher than those found by Seliger *et al.*, in 1970 (139 and 151 bpm). The highest value was obtained on the high bar (170 ± 2 bpm); then, in descending order, parallel bars (158 ± 4 bpm); rings (149 ± 5 bpm); pommel horse (145 ± 7 bpm); and vault (130 ± 4 bpm). It was not until 1982 that Montgomery and Beaudin assessed the HR in the four female routines using telemetric recording. The peak HR was much higher than those found by the previous investigations: 178 ± 11 bpm. The average HR was 166 ± 10 bpm. Later in the 90s, Goswami and Gupta (1998) studied the cardiac responses during full male routines in separate sessions using heart rate monitors (Sport Tester PE-3000). The peak HR of five events (floor, pommels, vault, parallel bars and horizontal bar) was 180 ± 5 bpm and the average was 161 ± 9 bpm. These values are much higher than those found in the 70s, although the recording was performed with gymnasts practising at a similar level (non-elite). This confirms the increasing difficulties of the gymnastic elements.

Le Chevalier *et al.* (1999) recorded the cardiac response during three sessions of different intensities, using a pulse-by-pulse heart rate monitor (BHL Bauman 6000):

- an 'easy session' based on skill rehearsal on the floor, pommels and rings;
- a competition in which the gymnasts performed their six routines;
- a double competition, in which the gymnasts performed their six routines twice (called also 'back-to-back').

During the competition and the back-to-back sessions, gymnasts' HR max was close to their HR max measured in a maximal incremental test (average HR 122 ± 7 bpm and 124 ± 8 bpm respectively). However, their HRs were around 60% of the HR max during the skill-rehearsing session and therefore significantly lower than the two other sessions (average HR 114 ± 10 bpm).

**TABLE 2.1** Heart rates in bpm during gymnastic exercises

| *Females* | *Seliger (1970)* | *Noble (1975)* | *Montgomery (1982)* | *Viana & Lebre (2005)* | |
|---|---|---|---|---|---|
| Vault | 133 ± 13 | – | 162 | – | |
| Un bars | 133 ± 10 | 167 ± 2 | 187 | 195 ± 10 | |
| B beam | 130 ± 13 | 159 ± 6 | 177 | 179 ± 8 | |
| Floor exe | – | 169 ± 6 | 185 | 193 ± 2 | |
| Max | 148 | 189 | 187 | 205 | |
| Average | 132 | 165 | 178 | 189 | |
| *Males* | *Goswami & Gupta (1998)* | *Lechevalier et al. (1999)* | *Groussard et al. (2002)* | *Jemni et al. (2000, 2002, 2003)* | *Viana & Lebre (2005)* |
| Floor exe | 183 ± 11 | 186 ± 5 | 160–179 | 186 ± 11 | 182 + 2 |
| Pommels | 173 ± 9 | 188 ± 7 | 158–170 | 185 ± 11 | 174 + 8 |
| Rings | 175 ± 10 | 188 ± 6 | 156–165 | – | 171 ± 15 |
| Vault | – | 160 ± 9 | 160–179 | 162 ± 14 | – |
| Parallel b | 175 ± 15 | 183 ± 7 | 154–162 | 181 ± 11 | 176 + 5 |
| High b | 182 ± 12 | 187 ± 7 | 164–180 | 185 ± 9 | 183 |
| Max | 195 | 186 | 180 | 180 | 190 |
| Average | 178 | 181 | 160 | 167 | 177 |

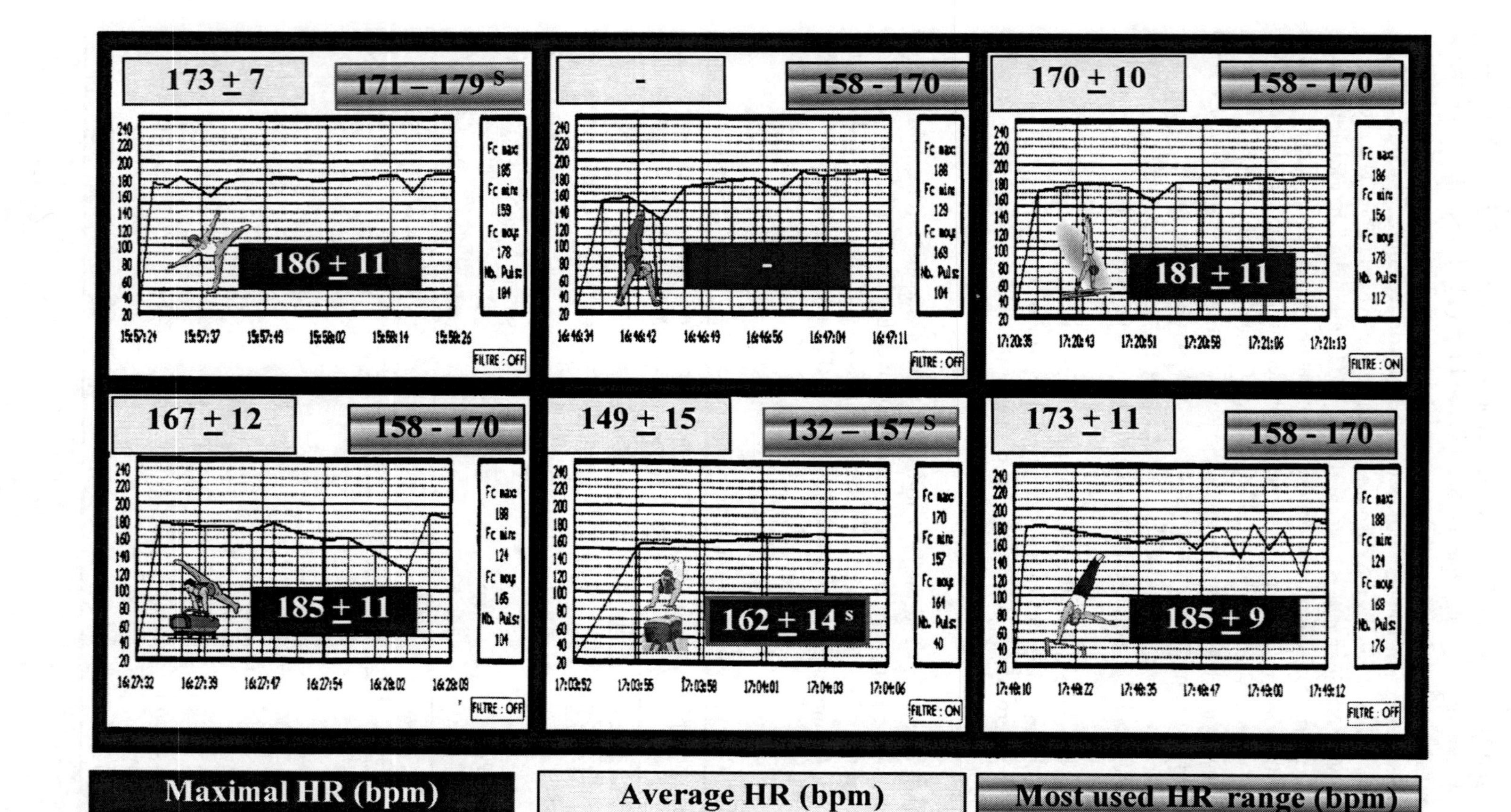


**FIGURE 2.1** Cardiac response during high-level male gymnastics competition (Jemni, 2001).

Figure 2.1 shows different recordings of the cardiac responses during a high-level male competition. HR was recorded continuously with heart rate monitors (BHL Bauman 6000). The recording shows the maximal values reached, the average and the HR range that the gymnasts have mostly used during each event. The highest HRs were recorded in the floor exercises and the lowest were obtained in the vault; these were significantly lower than all the other recordings and confirm the low blood lactate values found in the same apparatus (see Section 1.3.2). It is interesting to notice that peak HR was reached at the end of the floor, pommel horse, parallel bars and horizontal bar routines. This is due to the fact that gymnasts finish their routines with a higher technical element (dismount). Jemni (2001) demonstrated that gymnasts work closer to their HR max only during very short sequences. Indeed, the HR range between 180 and 190 bpm was found only during 16 ± 9% of all the recordings. The gymnasts worked mostly in an HR range between 158 and 170 bpm (28% of all recordings).

Figure 2.1 also demonstrates that gymnasts do not reach a steady state while performing their routines, therefore confirming the fact that energy cost cannot be estimated on the basis of such recordings. It is also important to mention that the same gymnasts have reached their lactic threshold at around 170 ± 10 bpm. Therefore, it can be concluded that cardiac response of high-level gymnasts is close to their metabolic threshold during full competitive routines, meanwhile peaking from time to time at their maximal HRs. These findings help clarify the nature of the metabolic stress during full gymnastics routines. However, HR analysis should be interpreted with caution, because running on a treadmill is different from gymnastics. In addition, HRs could be influenced by several variables, such as, catecholamines levels and would therefore make the achievement of any steady states a quite hard task.

## 2.4 Conclusion

- More details still need to be investigated in order to draw a full picture of the respiratory and cardiovascular responses to gymnastics.
- Several gymnastics elements, particularly those involving upside-down positions, slow down ventilation.
- It has been suggested that the greater upper-body mass of the gymnasts, when compared to similar age groups, may impinge upon the thorax and reduce its resting end-expiratory dimensions and hence the expiratory reserve volume of the lungs.
- There is evidence that gymnastics practice is not a sufficiently significant stimulus to enhance several heart variables including myocardial mass, cardiac output, heart rate, systolic volume and blood pressure.
- Some modifications of the local blood flow occur due to the centrifugal and/or centripetal force while performing several elements. However, it is quickly re-established at normal levels after the exercise.
- The development of the measurement tools from the sixties to the present allowed more reliable results and show an increased cardiovascular stress in parallel with the increasing technical requirements.
- Heart rates of high-level gymnasts are close to their metabolic threshold values during full competitive routines. Maximal heart rates are only reached for short peak periods.
- Gymnasts' heart rate recordings did not show any steady states. This makes them difficult to interpret for energy cost.

# 3

# FITNESS MODEL OF HIGH-LEVEL GYMNASTS

*William A. Sands*

## 3.1 The fitness model

Gymnastics physical fitness implies that there are some specific biological/physiological/physical regulatory processes that are at least somewhat unique to the gymnast. Indeed, there are specific physiological functions that are unique in terms of magnitude, timing and dominance in all sporting activities. In general, we describe a gymnast's physiological condition as his or her 'fitness'. This term implies that the body must be prepared, enhanced and maintained in order to perform gymnastics with competence. In short, fitness is 'readiness' for some type of activity.

Interestingly, no athlete can be 100% fit in all characteristics of physiology at the same time (Sands, 1994a; Sands *et al.*, 2001b). The body naturally economizes its adaptations to the demands that are imposed on it by training and performance. In fact, this economic adaptation has an acronym that accompanies it: SAID – specific adaptations to imposed demands. SAID means that the body will adapt to the demands made on it, no more and no less. A corollary to the SAID principle is that training must provide the athlete's body with an 'unambiguous message' of what you want the body to become.

> There are limits to the capacity of the athlete which are determined by his stage of development. During the competitive season (competitive period), the load tolerance and adaptability of the athlete is taxed by intensive competition and maximum workload at competition levels which take him to his limits. This leads to a particularly rapid development of the standard of performance; the previously developed bases are transferred to athletic performance, thus establishing optimum relationships between the performance factors, which lead to a definitive performance structure.
>
> Practical experience shows that this process cannot continue steadily in a linear fashion. It must be assumed that the load tolerance and the adaptation processes are so highly strained by threshold sustained activity that restrictions in activity in certain biological systems appear and adaptations which have not yet fully stabilized are temporarily lost.
>
> (Harre, 1982, p. 78)

The unambiguous message means that training must proceed in certain ways and that training for gymnastics will not be like training for other sports (Jemni *et al.*, 2000; Jemni *et al.*, 2006; Sands, 1991b; Sands *et al.*, 2003).

> The varied tasks of training cannot all be worked on at the same time. Care should be taken that a specific and systematic arrangement of immediate, intermediate, and long-term goals is made. Thus, observing the complexity of the main tasks and the continuous influence on all performance determining abilities, skills, and qualities of the athlete, certain areas of the standard of performance have to be emphasized for limited periods while others are simultaneously only stabilized or maintained.
>
> (Harre, 1982, p. 79)

For example, gymnastics training and performance require considerable strength fitness. Strength fitness is more effectively and efficiently acquired over the long term by beginning with strength-endurance, followed by maximal strength, then a focusing on the maximal strength to gymnastics-specific strength, and completed by a maintenance period that matches the duration of competitions. The aforementioned progression and development of sport-specific strength is a 'model' of training.

Models are simplifications of complex things such that the universe of things and ideas that could be taken into account is reduced to a more manageable number so that we can gain an understanding more rapidly and easily (Banister, 1991; Sands, 1993a, 1995). However, it is important to remember that models are not the 'real thing' and are only as good as the assumptions and constraints that are used to build them. With these cautions in mind let us look at a fitness model for gymnastics.

Fitness models are also sometimes called 'Profiles'. A number of investigations have looked specifically at determining a fitness model or profile for gymnasts (Sands, 1994a; Sands *et al.*, 2001a, 2001b). These models attempt to determine a sort of 'recipe' to determine the relative contributions of various fitness components for gymnastics performance. The model of Siff is particularly helpful as a starting point for fitness models for all sports (Siff, 2000). A modified model of gymnastics fitness consists of the following categories of fitness (Figure 3.1):

1. strength
2. speed
3. flexibility
4. skill
5. stamina (commonly used to mean 'muscular endurance').

Body composition can also be considered as the result of the interaction between the five components of the above model because gymnasts must lift their body weight against gravity. Body composition demands for gymnasts include a premium on leanness, but the idea probably doesn't rise to a level that merits a sixth fitness category. The gymnast's job is made considerably easier if he or she is light as well as strong (Malina, 1996; Sands, 2003; Sands *et al.*, 2002; Sands *et al.*, 1995). Moreover, gymnastics is an 'appearance sport', which means that the attractiveness of the athlete's body and performance is relevant. Anthropometric characteristics of gymnasts have been shown to be related to scores, particularly among female gymnasts (Claessens *et al.*, 1999; Pool *et al.*, 1969).

Each of these physical fitness categories is connected to every other category. We simply don't have the words in English to describe the combinations of physical fitness categories

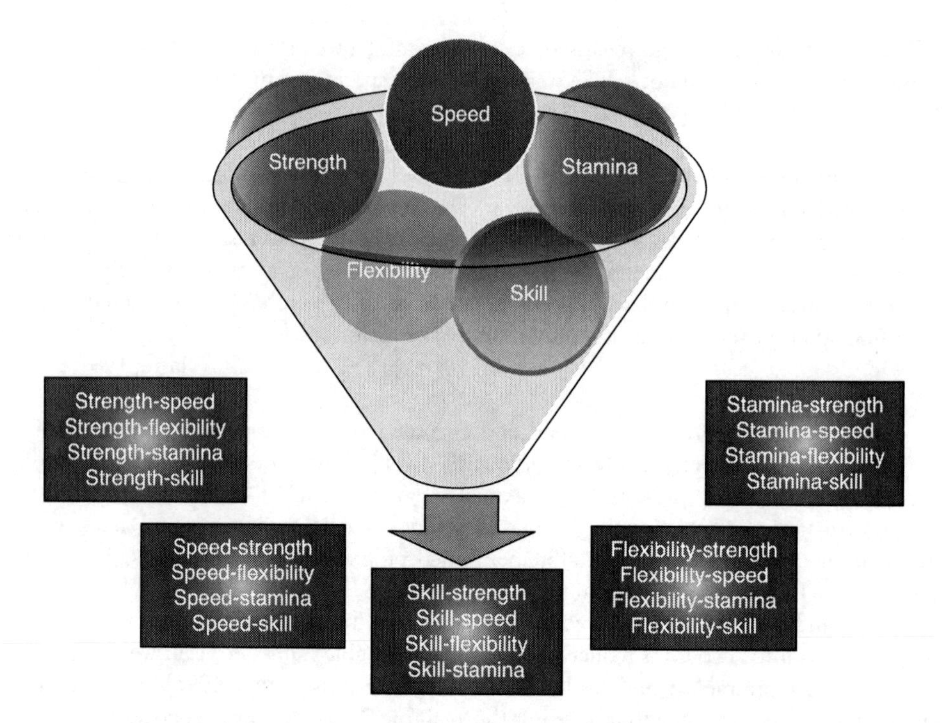


**FIGURE 3.1** Gymnastics physical fitness abilities (adapted from Sands, 2004). Note that strength, speed, skills, stamina and flexibility are considered as being equal to each other. Stamina is used to mean 'muscular endurance'.

that often spell the qualities of sport-specific fitness. For example, a fitness characteristic composed of strength and speed refers to the application of high levels of force rapidly. Speed-strength is a similar idea, but its emphasis is on the speed component. Thus, after combining all of the five categories, there should be 20 pairs of fitness components, as shown in Figure 3.1.

The combinations of fitness characteristics show how complex a fitness model might become; and we are not going to delve deeper into the three-way types of fitness, such as speed–strength–flexibility, and so on. The combinatorial explosion evident here is why training the athlete is so complex. Moreover, within this model are psychological characteristics of the athlete, inherited characteristics, including his or her genetic makeup and so forth. In other words, while the model helps to narrow the universe of things that merit consideration, we must acknowledge that actually getting the model to work in the real world with real people can be considerably more difficult (Sands, 1994a, 2000c; Sands *et al.*, 2001b).

## 3.2 Strength, speed, power, flexibility, stamina and skills

While acknowledging the interactions described above, the components of the simplest model of five fitness categories deserve more definitive individual treatment.

- *Strength* refers to the amount of force that can be exerted, under some pre-defined circumstances (Knuttgen & Komi, 1992; Sale & Norman, 1982; Wilk, 1990). It can be expressed in different modes: maximal/absolute, usually expressed in a single all-out effort; static or isometric; dynamic (slow, fast, concentric, and eccentric).

- *Speed* refers to the rate at which motions are performed (Joch, 1990; Mero, 1998; Verkhoshansky, 1996).
- *Flexibility* is usually defined as the range of motion in a joint or related series of joints such as the spine (Alter, 2004; Cureton, 1941; Holt *et al.*, 1995).
- *Stamina* is defined as the ability to persist in some motion in predefined circumstances for a predefined period. The ability to persist can come from several sources and is thereby one of the more complex aspects of athlete fitness. In the gymnastics context, stamina refers mainly to muscular endurance. For example, the gymnast can enhance his or her ability to persist by increasing metabolic efficiency and effectiveness via training of specific energy system enzymes, substrates, mediators and pathways (Sale & Norman, 1982; Sands, 1985; Sands *et al.*, 2001a). The gymnast can also enhance his or her stamina for short-term endurance (task demands under two minutes) by increasing maximal strength (Jemni, 2001; Jemni *et al.*, 2006; Sands *et al.*, 1987; Sands *et al.*, 2001a; Sands *et al.*, 2004a; Stone *et al.* 1984).
- *Skill* is the coordinated application of forces, positions and movements to accomplish a predefined task (Abernethy *et al.*, 1998; Schmidt & Young, 1991).

In addition to the model above is the concept of body composition. A gymnast must be 'on the lean side of lean' because he or she must lift his or her body mass against gravity through complex and difficult positions while maintaining exquisite control.

## 3.3 Conclusion

- Strength, speed, flexibility, skill and stamina (or muscular endurance) are the main physical abilities of gymnasts. Body composition and coordination are considered as prerequisite at high level.
- Modern artistic gymnastics stresses the importance of 'strength' and 'speed', or in other words: power.
- All these physical abilities interact, showing 20 pairs of fitness components essential to build up a physical fitness model for gymnasts.

# 4

# TRAINING PRINCIPLES IN GYMNASTICS

*Monèm Jemni and William A. Sands*

Different characteristics of fitness are combined like a recipe to result in optimal fitness for gymnastics. However, the recipe for optimal fitness also demands several other concepts. These are the principles of training:

## 4.1 Specificity

Although specificity was briefly described above, it is important to appreciate just how specific training and testing are in the development and monitoring of a gymnast. Early training of young athletes or the early stages of a long preparatory period may involve more 'general' training, also called 'multilateral development' (Bompa & Haff, 2009). However, specificity remains one of the most important principles involved in athlete training. A gymnast trains by performing on the apparatus, not by swimming or running. Although the gymnasts may get some training benefits from swimming and running, specificity garners the greatest return on training investment.

Training and testing are specific to the position of the body (Behm, 1995; Oda & Moritani, 1994; Sale & MacDougall, 1981; Sale, 1986), to the speed of movement (Moffroid & Whipple, 1970; Sale & MacDougall, 1981), to the range of motion of the movement (Sale & MacDougall, 1981; Sale, 1992; Siff, 2000), to the sport (Müller *et al.*, 1999; Yoshida *et al.*, 1990), type of tension (Jurimae & Abernethy, 1997; Oda & Moritani, 1994), gender (Drabik, 1996; Mayhew & Salm 1990), and to the limb or side involved (Hellebrandt *et al.*, 1947; Sale, 1986). As such, the coaching and training of gymnasts should always consider the specificity of the exercises so that maximum transfer of training to gymnastics performance is more likely to occur.

## 4.2 Readiness

The principle of readiness refers to the idea that the road to competent performance in gymnastics is long and the path is seldom simple. Predictability of training exercises and loads is not guaranteed and the transfer of training approaches to actual performance is poorly understood (Bondarchuk, 2007; Christina & Davis, 1990; Sands *et al.*, 2003). However, readiness for performance has been addressed from a practical standpoint with the basic idea involving the questions a coach should ask and answer prior to teaching a skill or allowing a gymnast to perform a skill unaided (Sands, 1990a, 1990b). Clearly, ignoring the simple-to-complex approach to training and the physical and psychological competence of the gymnast leads to poor adaptation at best and injury at worst. The gymnast's readiness will develop with age and maturation. A gymnast at six years of age will benefit very little from anaerobic training. His or her body is not ready for conditioning at high intensities. However, an older gymnast is more 'receptive' to strength and conditioning/anaerobic training and can develop them more quickly.

## 4.3 Individualization

This training principle is commonly listed among the most important, in that each athlete has individual strengths and weaknesses and thereby needs or deserves an individualized training programme (Bompa & Haff, 2009; Sands, 1984; Stone *et al.*, 2007). It is obvious that gymnasts respond differently to the same training. This might be the result of differences in heredity, maturity, nutrition, rest, sleep, level of fitness, illness/injury, motivation and environmental influences. A wise coach should detect individual responses and formulate appropriate reactions for each athlete.

However, there are also circumstances when all athletes can and should do the same things. In many cases, coaching a team makes individualizing every training session a practical impossibility. Moreover, there are aspects of both development and enjoyment that profit from a team-based approach (Gould *et al.*, 1998; Hanin & Hanina, 2009; Loehr, 1983; Martin, 2002; Ravizza, 2002). The need for individualization of training increases as the athlete progresses in development. In the first few years of training, the young athletes all need to learn a large but complete list of basic skills (Sands, 1981). As the athlete reaches a higher standard, he or she is often using different skills and has clearly expressed his or her training and performance strengths and weaknesses. Fortunately, the size of the group or team of athletes is usually smaller in the latter stages of the athletes' careers due to simple attrition (Sands, 1995; Stone *et al.*, 2007).

## 4.4 Variation

Variation not only applies largely to the conditioning aspects of preparation, but also refers to the simple problem of boredom with training tasks and loads that remain fixed for periods beyond their effectiveness in causing adaptation (Arce *et al.*, 1990; Stone *et al.*, 2007). Training theory has placed a high premium on training variation due to the observation that unidimensional approaches to training do not appear to result in continued adaptations (Verkhoshansky, 1981, 1985, 1998, 2006). The general rule of thumb for training adaptation is that something about the training, such as the tasks, loads, durations, frequencies, etc., should change approximately every two to four weeks (Bompa & Haff, 2009; Verkhoshansky, 1981, 1998). Moreover, variation applies to training on a variety of levels. Athletes should experience varying training demands, varying skill levels, varying competition demands, varying yearly plans and varying levels of opponents. However, more recent information questions the idea of changing training loads so quickly, advocating training load changes only after approximately six weeks. Indeed, coaches often ask the following questions: Why wait six weeks before introducing a new training load? Why not change the training plan after two weeks when we notice that adaptations have occurred? Olbrecht (2000, p. 7) has answered these questions: 'The reason is that weeks 3 to 6 are necessary to stabilize the adaptations brought about in weeks 1 and 2'.

## 4.5 Diminishing returns

Related to the idea of training variation is the observation that a training programme and training stimuli suffer 'wear' (Harre, 1982). A training programme applied this year will not produce the same results if applied next year. Moreover, the early adaptations to training tasks are typically neural and occur rapidly while later changes (weeks or months later) involve structural changes (Moritani & DeVries, 1979; Sands & Stone, 2006; Stone *et al.*, 2007). From an evolutionary perspective, this approach to organism adaptation is 'smart': using neural adaptation first (i.e., learning) is not as demanding on available calories as structural changes. A novice gymnast would expect to learn new skills and gain some fitness at a much faster rate than those who have already been training for a number of years. However, most of the changes sought in athlete development require months or years to achieve and thereby require an enormous caloric/structural investment by the body. Wolff's law states that 'function determines structure' and there is no better example of this than in athletic training (Alter, 2004). The demands of training alter the functional demands placed on the body, which in turn results in modifications to structure. One author has observed that to achieve Olympic podium performance, the athlete must improve through approximately 20 preparatory and competitive periods (Bondarchuk, 2007). Fifteen periods of improvement usually lead to a high national and low international performance level. Ten periods of improvement lead to a national level of performance. Five periods of improvement usually result in competitive prowess at only the regional level. If an athlete cannot continue to improve for at least ten preparatory and competitive periods then he or she is likely to be untalented or to have chosen the wrong sport (Bondarchuk, 2007). A related observation regarding diminishing returns is that improvement can be represented by a decelerating curve. Improvements are rapid in the beginning and slow down dramatically or plateau as the athlete reaches his or her genetic ceiling (Sale, 1992).

## 4.6 Regeneration and the new concept of recovery in gymnastics

Training should be considered a unity of both training load and recovery-adaptation. Recovery does not start after the training session, it actually starts within the session if the right 'duration and method' are imposed. However, gymnastics coaches seldom question the effectiveness of their recovery 'procedure'. Gymnasts rarely adopt 'active types of recovery' compared to other sports, where it has shown substantial benefits (Dodd *et al.*, 1984; Stamford *et al.*, 1981). It is now understood that 'overtraining could be avoided if recovery/regeneration is planned effectively'.

> Establishing the unit of work and recovery is essential for the effectiveness of athletic training. Training causes fatigue which occasions a temporary lowering of performance. Hard work is followed by two processes which have already been introduced during the work itself and which may temporarily run along parallel lines:
>
> - the recovery process leading to the re-establishment of the full ability to function, and
> - the adjustment processes leading to the functional improvement of performance and the morphological reorganization of the functional systems under stress.
>
> (Harre, 1982, p. 65–66)

Once recovered, the body will enter a period of overcompensation where the body's systems have adapted beyond the original threshold. Optimal post-training recovery allows the athletes to take part in their other daily activities, such as studying, working or socializing. Harre (1982) recognized that overloading is not actually possible unless the athletes have recovered. The quicker the recovery, the faster the shift to the new training stimulus. Moreover, if the recovery is repeatedly insufficient, fatigue builds up, performance deteriorates and there is increased risk of injuries.

Gymnastics competition is a series of performances on several kinds of apparatus interspersed with rest periods. It is in fact similar to circuit training, owing to the intermittent and intense activities that are involved. It is therefore important to design the optimal means of recovery so that the gymnast can begin each event without undue fatigue. Fatigue is indeed more than a simple physiological problem; it may lead to falls and injuries in gymnastics (Sands, 2000a; Sands, 2000b; Tesch, 1980).

During intense sessions, gymnasts are asked to perform routines while fatigued (Jemni *et al.*, 2000). They are usually asked to repeat their events several times per practice session, which leads to a high level of lactate production and accumulation (Le Chevalier *et al.*, 1999). They are therefore required to find the best compromise between technical effectiveness, safety and high intensity effort. Jemni *et al.* (2003) have set new recovery guidelines that could assist gymnasts in reducing their blood lactate between competitive events and could therefore enhance regeneration. They have shown that if a combined period of rest and active recovery, where heart rate is kept below the anaerobic threshold, is incorporated between the events, it not only enhances lactate clearance but also helps the subsequent performance. Blood lactate clearance was indeed significantly higher when using the combined passive/active recovery when compared to the 'classical passive only' (40.51% v 28.76%, respectively).

To conclude, training plants the seeds of performance, careful nurturing of the seeds allows them to sprout (periodization) and recovery-adaptation makes the plants grow and bear fruit. Gymnastics training should include periods of regeneration such that the athlete is fully prepared and not fatigued when a new training challenge is posed.

## 4.7 Overload and progression

The gymnast must attempt tasks that are initially beyond his or her capability in order to force the organism to fatigue and later recover and adapt. Overload is the term used to describe the task(s) that the gymnast performs that exceed his or her current performance limits. Progression in gymnastics is a key principle; it refers to the idea that skills should be learned in a systematic progression, starting with the most basic set of skills and evolving to the more complex. This not only allows the building up of the 'technical repertoire' of the gymnasts step by step, but in particular avoids systematic errors which are tough to correct at a late stage. A current example of this is a gymnast who struggles to perform a correct 'round-off'; in most cases it would be revealed that this gymnast has 'jumped' a learning stage, obviously the 'cartwheel'.

In addition, there are limits to the amount of overload an athlete can withstand without injury, breakdown and excessive fatigue. The ideal training programme provides optimal challenges that the athlete can barely handle and then the athlete rests (reduces training demands) to promote recovery-adaptation. Too high and/or too challenging loads may affect the motivation of the athlete and may lead to loss of interest. Meanwhile, too low a load/challenge does not allow any benefits. The athlete is rewarded during recovery-adaptation with new skills and abilities that were caused by the previous overload.

Thus, training is an undulating series of training challenges given in planned, successively increasing loads (i.e., overload) taking into consideration the initial state (i.e., progression) and allowing a shift to a higher level, allowing supercompensation in order to achieve training goals (Sands, 1984, 1987, 1991a, 1994b; Stone *et al.*, 2007). Overload is multi-dimensional, including all aspects of psychology and fitness. Maintaining order and a systems approach requires a specialized methodology. That methodology is called 'periodization'.

## 4.8 Periodization

All of the previous training principles are often included or embodied in the concept of periodization.

> Periodization is the division of a training year into manageable phases with the objective of improving performance for a peak(s) at a predetermined time(s).
>
> (Smith, 2003, p. 1114)

Periodization involves two simultaneous concepts: cyclic variation in training load and the division of training demands into separate phases (i.e., periods) (Bompa & Haff, 2009; Harre, 1982; Matveyev, 1977; Verkhoshansky, 1985). Periodization is currently a very confused term with a variety of investigators, coaches and administrators weighing in on exactly what it means, what concepts are most important and how to implement the ideas.

Periodization begins by developing an annual plan that is composed of one or more *macrocycles* that include a preparatory period, a competitive period and a transition period. The macrocycle is broken down into *mesocycles* that are usually 4 to 6 weeks in duration; the total number of training demands are compartmentalized in such a way that only a few are emphasized during each mesocycle. Each mesocycle has its own goals and each mesocycle builds on previous mesocycles. Within each mesocycle are *microcycles*. Microcycles are approximately one week in duration and involve a more focused group of tasks and goals that function to achieve an accumulation of training stimuli that will force the athlete to experience the unity of an overload and recovery-adaptation. Interestingly, it takes about a week of continuous load demands to cause a training stimulus sufficiently large that the athlete's body will show the effects of overload (i.e., fatigue), with recovery and adaptation following the overload.

Linking microcycles and mesocycles together in a systematic way rewards the coach and athlete with improved training and performance ability. The entire 'map' of this development is called periodization. There are many periodization models involving the controlled application of *volume* (how much the athlete does, which might be expressed in terms of time or number) and with *intensity* (how hard the challenges are, which might be expressed in number per unit time) to achieve a balanced training experience that leads to progress without injury or the threat of overtraining. Of course, periodization tenets also attempt to ensure that the athlete maximizes his or her abilities, so that the athlete is not defeated by opponents who simply work harder.

Planning a training cycle in high-level gymnastics should take into consideration the two peaks of the season: one for the continental Championships (such as the European Championships), usually between April and June, and the second for the World Championships or Cup, around October/November. Therefore, there are two macrocycles, each one lasting approximately six months and composed of the three periods, i.e. preparation, competition and transition. Each period is composed of mesocycles with specific objectives, and these in turn are divided into weekly microcycles. Members of the national squads are always gathered into regular training camps and pursue one or two training sessions per week in the regional centre in order to achieve the objectives set.

Arkaev & Suchilin (2004) showed that the structure of an Olympic cycle has to take into consideration the four annual training cycles. The whole cycle would be composed of eight six-month macrocycles. A specific objective is set for each macrocycle, starting by raising the level of specific physical training and ending by the final selection of the Olympic team a few months before the games. Coaches and technical staff have to plan ahead and consider all the following with their plan: physical/functional preparation; monitoring; tactical preparation; psychological preparation; jump preparation; acrobatic preparation; choreographic preparation; camp preparation; friendly competitions; and medical preparation.

## 4.9 Conclusion

- As in all other sports, gymnastics coaching is based on the main generic training principles, which are: specificity, readiness, individualization, variation, diminishing returns, overload and progression, periodization and regeneration. Applying science to gymnastics has allowed this sport to evolve by adopting new concepts in training and coaching, such as the new concept of recovery in gymnastics.
- Long-term planning facilitates the achievement of the objectives via a structured periodization which takes into account all the above principles.

# 5

# SPECIFIC PHYSICAL AND PHYSIOLOGICAL ASSESSMENTS OF GYMNASTS

*Monèm Jemni*

The complexity of gymnastics events (six for males and four for females) requires not only different training approaches, but also a wide range of physical and physiological testing in order to monitor the progress of each gymnast. However, the measurements are made difficult by having a number of complex parameters, such as a wide variety of technical skills, muscular contractions and speed of stretch but only a limited number of standardized specific tests.

This section highlights some of the tests most widely used in assessing the physical ability groups.

## 5.1 Strength and power tests for upper and lower body

### *5.1.1 Standardized laboratory tests*

The standardized laboratory tests that have been applied in gymnastics include maximal oxygen uptake protocols for upper and lower body, force–velocity tests and Wingate tests. These are widely accepted, valid and reliable tests in sport science, often considered to be the gold standard for assessing aerobic and anaerobic metabolisms. Results of such tests performed by male and female gymnasts of different levels are presented in Section 1.3.

### *5.1.2 Specific jumping and plyometric tests*

The literature provides a variety of jumping tests which might be used in different sports contexts (Fry *et al.*, 2006; Gabbett, 2006; Lidor *et al.*, 2007; Melrose *et al.*, 2007). All assess the height of the jump (and/or the displacement of the centre of mass) and indirectly the power output via different equations using mainly body mass. Some authors have estimated the power output by using contact mats and therefore have highlighted other indices, such as flight and contact times (Bosco *et al.*, 1983; Loko *et al.*, 2000; Markovic *et al.*, 2004; Sipila *et al.*, 2004). The force plate, however, gives better measurement of the power output developed during the jump (Carlock *et al.*, 2004). Nevertheless, the most accurate jumping profile might be set, including kinetic and cinematic analysis, especially if the force plate is synchronized with video capture/analysis. All authors agree on the strong relation between the height of the jump and the power output (Van Praagh & Dore, 2002). Other components, such as

**TABLE 5.1** Jumping test results for male and female gymnasts

| | *Jumping tests* | | | |
|---|---|---|---|---|
| | *Females* | | *Males* | |
| | *Results (cm)* | *Norms for high level* (cm)* | *Results (cm)* | *Norms for high level* (cm)* |
| Vertical jump with arms swing | 47.8 (Sands 1993)<br>45.2 (Marina 2003)<br>49.2 (Heller *et al.* 1998)<br>42.5 (Bale *et al.* 1987) | 50–60 | –<br>52.8 (Marina 2003)<br>53.9 ± 10.9 (Jankarik *et al.* 1987) | 60–70 |
| Vertical jump without arms swing | 38.2 (Marina 2003) | 40–45 | 41.5 (Marina 2003)<br>41.3 ± 2.3 (Leon-Prados, 2006) | 50–55 |
| Standing long jump | | 210–230 | | 225–245 |
| Drop jump | 45.6 (from 60 cm) (Marina 2003) | | 50.9 (from 60 cm) (Marina 2003)<br>26.1 (Faria *et al.* 1989) | |
| Drop jump with jump off | | 60–65 | | 70–80 |
| Squat jump | | | 36.6 ± 2.3 cm (León-Prados, 2006) | |

*: Arkaev and Suchilin (2004).

the body's vertical velocity at takeoff, the peak jumping velocity and the impulse (force multiplied by time) are also very important to consider in studying jumping ability (Bobbert & Van Ingen Schenau, 1988; Haguenauer *et al.*, 2005; Winter, 2005). However, to maximize all of these factors, segmental coordination and a proper technique are essential in order to obtain the best performance (Bobbert, 1990; Carlock *et al.*, 2004). As a matter of fact, it is widely accepted that analysing jumping performance should take into consideration more than one factor, in particular when comparing subjects (Marina *et al.*, 2011).

In all cases, the calculated power output does not reflect the physiological mechanisms underpinning the athletes' performance. Jumping tests have been widely accepted and are applied in gymnastics (Table 5.1). It has indeed been shown that the use of the jumping mat is an effective assessment tool for gymnasts, which simulates training and allows monitoring (Sands *et al.*, 2004). Bouncing and jumping indeed make up important parts of the floor, balance beam and vault routines. Gymnasts learn this skill at the very early age of specialization as part of their daily training. It allows take-off skills and high aerial acrobatics skills. Table 5.1 shows results of some jumping-test investigations, as well as norms for high-level males and females.

Plyometric work with the upper and lower body is extensively used with all gymnastic apparatus. Gymnasts are very often required to combine an aerial element with another as soon as they land. In addition, all catches and releases of the apparatus require plyometric contractions, while maintaining a straight body line. Gymnasts learn and develop this quality at a very early age in order to be able to maintain the basic handstand. Marina *et al.* (2011) have compared plyometric performances of elite male and female gymnasts with similar age- and gender-matched groups (Figure 5.1). Elite gymnasts obtained significantly better

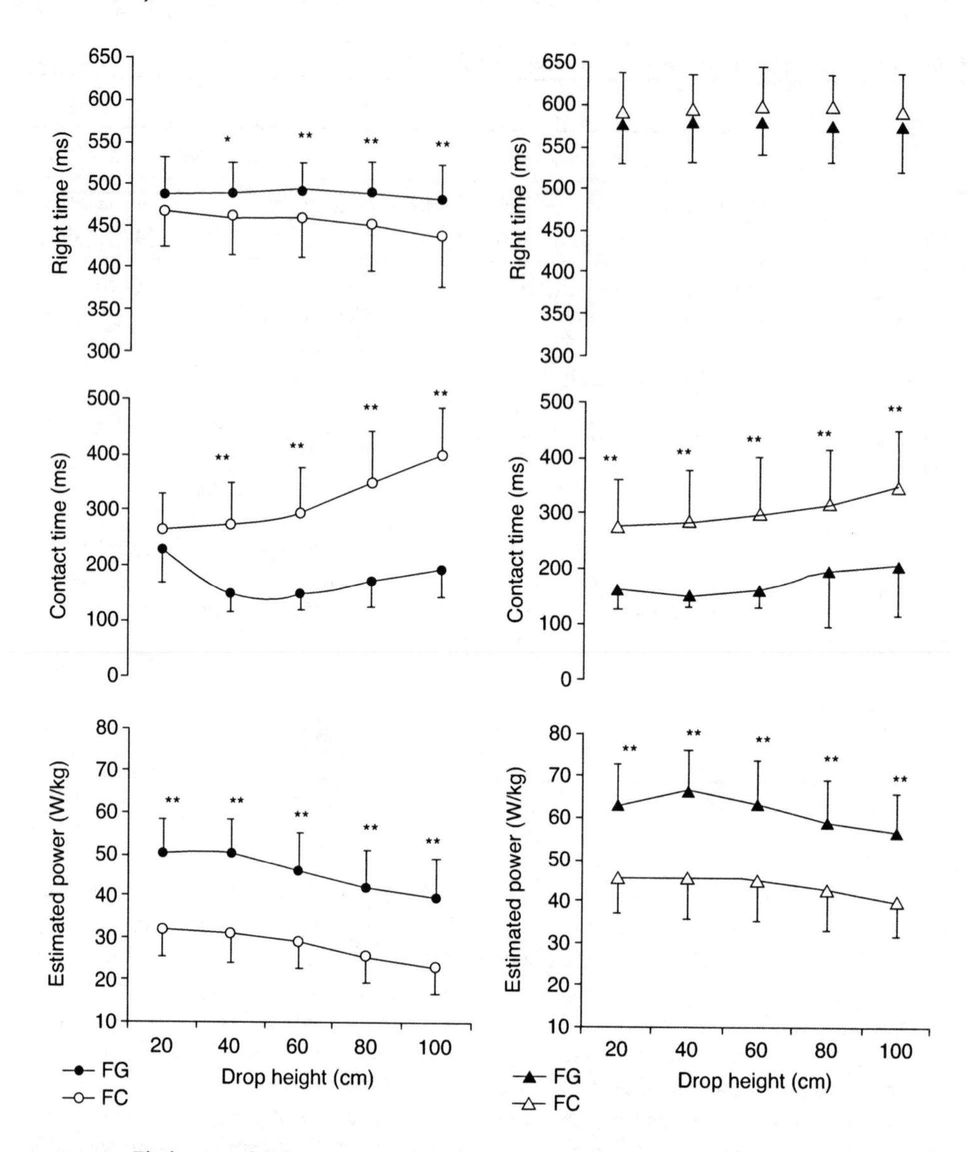


**FIGURE 5.1** Flight time (FT), contact time (CT) and estimated power (W/kg) during plyometric jumps from 20, 40, 60, 80 and 100 cm.

FG: Female Gymnasts; FC: Female Control; MG: Male Gymnasts; MC: Male Control. *: $p \leq 0.01$. **: $p \leq 0.001$. [Adapted from (Marina *et al.*, 2011).]

performances at drop heights from 20 to 100 cm. Most of the male and female gymnasts obtained their best performance between 40 and 60 cm, with the exception of the best elite gymnasts of both genders, who achieved their best at 80 cm.

### 5.1.3 Muscular endurance tests

Muscular endurance is a significant discriminator of high- versus low-level gymnasts (Sands, 2003). It is the ability to perform for extended sequences, such as long routines, without

**TABLE 5.2** Muscular endurance tests for gymnasts

| | | *Muscular endurance tests* | |
|---|---|---|---|
| | | *Females* | *Males* |
| Bosco 60-sec jumps | (Sands *et al.* 2001b) | 23.7 W/kg | – |
| Max leg lifts | (Lindner *et al.* 1992) | 13.2 ± 3.8 | – |
| Max pull-ups | (Seck *et al.* 2005) | – | 16.6 ± 4 |
| Push-ups in 60-sec | (Grabiner *et al.* 1987) | – | 292.5 ± 45.8 W/kg |
| Sit-ups in 60-sec | (Grabiner *et al.* 1987) | – | 201.1 ± 28.5 W/kg |
| Dips in 60-sec | (Grabiner *et al.* 1987) | – | 163.4 ± 41.1 W/kg |
| Pull-ups in 60-sec | (Grabiner *et al.* 1987) | – | 97.6 ± 31.4 W/kg |
| Support half-lever | (Arkaev *et al.* 2004) | 28–30 sec | – |

undue fatigue. Gymnasts develop this quality progressively with age and practice. It is habitually assessed through specific tests performed on the apparatus engaging upper and/or lower limbs or several muscular groups and/or joints at the same time. Table 5.2 shows examples of muscular endurance tests for male and female gymnasts. For example, the leg lifts test requires the gymnast to fully lift his or her straight legs together, up to 90 degrees (horizontal) while hanging at the espalier or the bar. This test has been used in different contexts, including talent identification and selection. The repeated jumping test is also commonly used for lower-body muscular endurance. One of the most used is the 60-second Bosco test, which involves repeated use of the stretch–shortening cycle of the lower extremity. Sands and colleagues have used this test to select the women's Olympic Gymnastics members during their seven months of trials leading to the Sydney Olympic Games (Sands, 2000b; Sands *et al.*, 2001a, 2001b). The average power outputs of these trials were as follows:

- Female US Senior National (*n* = 34, 17.2 years): 23.7 ± 5 W/kg
- Female US Senior National Team (*n* = 6, 17,3 years): 23.0 ± 4.8 W/kg
- Female US Junior National Team (*n* = 15, 13.9 years): 21.6 ± 2.8 W/kg.

### *5.1.4 Agility, speed, strength and power tests*

In order to acquire strength, agility, speed and power, gymnasts vary their training according to the specificities of the skills, as well as the specificity of the apparatus. While some types of apparatus require relatively slow-moving or held positions (i.e., isometric) and extraordinary strength, such as the cross and the Maltese at the rings, other events need explosive power development in very short sequences, such as tumbling. Gymnasts also need exceptional agility, together with remarkable spatial awareness in order to achieve very complex elements, including catches and releases of the apparatus after somersaulting and twisting. Agility, speed of movements and coordination between the hands and the rest of the body are also required to perform on the pommel horse.

Although speed could be assessed with several sprinting distances, gymnastics tests rarely exceed 20 m, which corresponds to the maximum distance allowed as a run-up for the vault. Otherwise, agility, strength and power are often assessed by specific tests on the apparatus involving hanging or support positions. One of the classical tests recognized worldwide is rope-climbing. Different climbing distances are set for males and females; in most cases the

**TABLE 5.3** Specific tests for strength, agility, speed and power in gymnastics

| *Strength, agility, speed and power tests* | | | |
|---|---|---|---|
| | | *Females* | *Males* |
| Rope climb without leg help | Arkaev *et al.* (2004)<br>León-Prados (2006) | (4m) 5.5 to 6 sec | (3m) 5 to 5.5 sec<br>(5m) 5.7 ± 1.3 sec |
| Full leg lifts in 10 sec from hanging position at espalier | Sands (1993) | 6.2 reps | – |
| Push-ups in 10 sec | Sands (1993) | 14.1 reps | – |
| Pull-ups in 10 sec | Sands (1993) | 7.2 reps | – |
| 20-m dash | Sands (1993)<br>Jankarik *et al.* (1987) | 3.1 sec | –<br>3.6 ± 0.5 sec |
| 30-sec jumps | León-Prados (2006) | | 30.4 ± 10.9 W/kg |

athletes are not allowed to use their feet, so the test is performed with bent knees or in half lever. Table 5.3 shows some data from examples of strength, agility, speed and power tests.

## 5.2 Flexibility tests

Gymnastics is a sport which requires a great range of motion in most joints. Flexibility plays a considerable role in the success of a routine. In many cases, the score is directly influenced by the possibilities of a gymnast's 'body motion'. Gymnasts are indeed marked on 'how perfectly they perform' and not on 'how many skills they present'. Full amplitudes of the movements are continuously required. A slightly bent ankle, for example, is penalized; therefore gymnasts learn to be 'over-flexible' in order to achieve the required range of motion in a natural fashion. It is very common to observe an over-split, while a gymnast is stretching or performing: the gymnast's legs are actually stretched above the horizontal position (i.e. with more than 180 degrees between them). In addition, lack of flexibility in one or more joints may slow down the learning process or make it quite difficult. Upper and lower limbs, neck and spine should demonstrate a variety of positions: forward, backward, sideways and sometimes in a longitudinal plane, as is the case for dislocation elements on the horizontal bar and rings. It has in fact been demonstrated that gymnasts are the most flexible athletes (Kirby *et al.*, 1981). However, it has also been shown that rhythmic gymnasts are more flexible than artistic gymnasts (Douda & Toktnakidis, 1997).

Flexibility assessment in gymnastics is a daily routine at the very beginning of the specialization stage when athletes are developing their basic physical skills. Once an optimal range of motion is gained, it will be maintained by daily stretching. Flexibility tests include passive and active forms, most commonly in different planes: forward, backward and sideways. Table 5.4 shows some of the tests performed in gymnastics. It is indeed important to maintain a symmetrical range of motion from both sides in order to perform the skills at full amplitudes. It has been shown that artistic female gymnasts have better symmetry than rhythmic gymnasts (Douda & Tokmakidis, 1997). This fact might be explained by more active flexibility being employed in artistic gymnastics, which is indirectly influenced by the level of strength of the agonist muscles. Personal interviews with the coaches of high-level rhythmic gymnasts revealed that gymnasts tend to put more emphasis on stretched skills with their leading leg because they have to handle the apparatus at the same time.

**TABLE 5.4** Flexibility tests in gymnastics

| | | *Flexibility tests* | | |
|---|---|---|---|---|
| | | | *Females* | *Males* |
| Passive flexibility | Sands 1993 | Right forward split | 35.5 cm | – |
| | Sands 1993 | Left forward split | 31.4 cm | – |
| | Jankarik *et al.*, 1987 | Hip flexion | – | 149,0 ± 76,9 degrees |
| | | Back extension | – | 97,3 ± 1,82 degrees |
| | | Shoulder extension | – | 47,2 ± 1,35 degrees |
| Active flexibility | Sands, 1993 | Shoulder flexion | 48 cm | – |
| | | Left forward leg lift | 7.3 pts* | – |
| | | Right forward leg lift | 8.6 pts* | – |
| | | Left sideward leg lift | 7.9 pts* | – |
| | | Right sideward leg lift | 8.9 pts* | – |
| | Jankarik *et al.*, 1987 | Hip flexion | – | 111.7 ± 52.4 degrees |
| | | Back extension | – | 71.1 ± 46.9 degrees |
| | | Shoulder extension | – | 27.4 ± 12.7 degrees |

*: ankle above chin = 1 0 pts; above shoulder = 9 pts; above chest = 8 pts; above hip = 6 pts; below hip = 3 pts.

Finally, increasing interest has been shown in the enhancement of the range of motion by the new 'vibration therapy'. Several studies have shown that this technology might be a promising means of increasing flexibility beyond that obtained with static stretching. Sands *et al.* (2006b) have not only increased the forward split amplitudes of a high-level group of male gymnasts when compared to a control group (both were already flexible) but also succeeded in saving time.

## 5.3 Technical tests

Technical tests have been applied in gymnastics since the very early days of the discipline. Technical tests allows not only checking the readiness of the gymnasts but also selection. Becoming a member of the national squad is mainly based on performance and ranking in competitions; however, several countries have set up a parallel stream for young talents who engage in a selection process based on physical and technical abilities, such as the 'Talent Opportunity Programme' (TOP) in the USA (Sands, 1993b). In addition, monitoring basic technical skills is a regular procedure at a higher level of performance. However, it appears that at the sub-elite level, coaches often wait for competitions to check their athletes' preparation. Table 5.5 shows some of the common technical tests in artistic gymnastics.

**TABLE 5.5** Technical tests in gymnastics

| | *Specific tests* | | |
|---|---|---|---|
| | | *Females* | *Males* |
| Handstand push-ups in 10 sec | Sands (1993) | 7.5 reps | – |
| Held handstand | Arkaev *et al.* (2004) | 9 sec | – |
| Handstand push-ups in 60 sec | Grabiner *et al.* (1987) | – | 38.8 ± 11.8 reps |

*(continued)*

**TABLE 5.5** (*Cont'd*)

| *Specific tests* | | | |
|---|---|---|---|
| | | *Females* | *Males* |
| Cross on rings | Arkaev *et al.* (2004) | – | 5–6 sec |
| | León–Prados (2006) | – | 6.7 ± 4.3 sec |
| Maltese on ring | León–Prados (2006) | – | 3.3 ± 1.8 sec |
| Front horizontal hang | Arkaev *et al.* (2004) | 20–23 sec | 5–6 sec |
| Back horizontal hang | | 28–32 sec | – |
| Horizontal support (planche) | Arkaev *et al.* (2004) | – | 5–6 sec |
| Inverted cross | Arkaev *et al.* (2004) | – | 5–6 sec |
| From support half-lever, lift to | Arkaev *et al.* (2004) | 8–10 reps | – |
| handstand with straight arms and bent body | León–Prados (2006) | – | 10.9 ± 2.9 reps |
| From swing on low bar, upstart | Arkaev *et al.* (2004) | 10–12 reps | – |
| to handstand and repeat | Sands (2000) | 10–15 rep | – |

## 5.4 Conclusion

A wide range of physical and physiological tests are applied in order to monitor the progress of the gymnasts. However, most of these tests are non-invasive, field-based and indirectly reflect the dominant metabolic pathway. The lack of specific laboratory tests for these athletes make authors speculate about the energy, force and power requirements. Investigators still have to come up with the 'gold-standard' tests for gymnastics. Meanwhile, the tests currently being applied are widely accepted, such as jumping and flexibility tests. The accuracy, reliability and validity of the specific technical tests are quite low because of a 'learning effect', and also because different coaches and scientists perform the test differently.

# 6

# DIET, NUTRITION, SUPPLEMENTATION AND RELATED HEALTH ISSUES IN GYMNASTICS

*Monèm Jemni*

## 6.1 Diet in male and female gymnasts

In gymnastics, as in all artistic sports, special consideration is given to external appearance. Female gymnasts and particularly rhythmic gymnasts suffer from several pressures: the most direct is the pressure of the coach who, in most cases, is directly concerned by the weight of the gymnasts; the pressure of the judges or referees, who give a mark on aesthetic abilities and appearance; and the pressure of putting on weight, which may reduce their acrobatic abilities. These pressures may push the gymnasts and the coaches towards severe dietary habits, which may lead to a negative energy balance (Rosen & Hough, 1988). It has been demonstrated that the risks of eating disorders, such as anorexia nervosa and bulimia, are increased in athletes performing sports which emphasize leanness (Brotherhood, 1984; Jankauskienė & Kardelis, 2005). Almost 100% of the reviewed articles agreed on the eating disorder among gymnasts. Evidence showed that such diets lead to a decreased performance and an increased risk of injuries, which might be associated to some severe health issues. Benardot (1999), for example, has attributed the incidence of stress fracture in rhythmic gymnastics to their negative energy balance and demonstrated that 86% of the high-level gymnasts had at least one serious injury per year.

For both genders, total energy consumption and nutritional intake are insufficient in spite of high energy expenditure, although to a lesser extent in male gymnasts (Filaire & Lac, 2002; Fogelholm *et al.*, 1995; Lindholm *et al.*, 1995). These athletes may practise more than six hours per day in certain periods, such as during training camps (Sands, 1990b; Stroescu *et al.*, 2001). Interviews with former rhythmic gymnasts showed that it is very common to practise for more than 10 hours per day. The energy intakes reported in Chinese sub-elite and elite female gymnasts were 1,637 kcal/day and 2,298 kcal/day respectively (Chen *et al.*, 1989). The authors demonstrated that both groups consume high percentages of fat compared to proteins (~43% v 13% respectively). However, this might be the result of lack of knowledge about dietary and healthy eating. Filaire & Lac (2002), for example, found different diet portions in French female gymnasts (14% protein, 48% carbohydrate and 37% fat). Nevertheless, both studies, as well as Lindholm *et al.* (1995), have demonstrated that female gymnasts have deficiencies in several minerals, in particular calcium, thiamine and riboflavin, iron, fibre and E and B6 vitamins, when compared to similar aged females.

**TABLE 6.1** Food sources for B complex vitamins according to the South African Gymnastics Federation (Humphy, 2010)

| *Vitamin* | *Best food sources* |
|---|---|
| Thiamine (B1) | Dried brewer's yeast, yeast extract, brown rice |
| Riboflavin (B2) | Yeast extract, dried brewer's yeast, liver, wheat germ, cheese and eggs |
| Pyridoxine (B6) | Enriched cereals, potatoes with skin, bananas, legumes, chicken, pork, beef, fish, sunflower seeds, spinach |
| Folate | Dried brewer's yeast, soya flour, wheat germ, wheat bran, nuts, liver and green leafy vegetables |
| Vitamin B12 | Liver, kidney, fish, red meat, pork, eggs and cheese |

In spite of the unbalanced diet and of the above deficiencies, it has been shown that protein intake in pre- and early pubertal female gymnasts was comparable to a control group (~7 g/kg per day) (Boisseau *et al.*, 2005). The authors have also shown that protein synthesis and degradation were also similar (~6 g/kg per day and ~5 g/kg per day, respectively).

However, male gymnasts have a more balanced diet. Some authors have even suggested that protein intake is quite high in order to enhance lean tissue formation (Brotherhood, 1984). Nutritionists recommend that gymnasts who train for more than eight hours per week should always consume nutrient rich foods such as vitamin-enriched and wholegrain cereals, fruits, vegetables and lean meats to ensure an adequate B-vitamin status. Total energy intake should also be adequate to maintain weight. Arkaev & Suchilin (2004) recommend a daily caloric intake between 4,500 and 5,000 kcal during intensive training periods for highlevel male gymnasts. Table 6.1 gives some generic indications of the most recommended nutrients

As a final point, there is an evident paucity of literature regarding male gymnasts' diet. However, female gymnasts' diet is more problematic and has been quite well investigated. Most of the investigations are in agreement and conclude that the unbalanced diet associated with their high energy expenditure could partly explain their low body mass index (BMI), small percentage of body fat, late pubertal development and irregular menstrual patterns, as explained in Sections 6.3.

## 6.2 Supplementation in gymnastics

Coaches always ask: do gymnasts need any dietary supplementation to perform, especially for high levels of strength and power?

Several investigations have confirmed that a significant percentage of elite gymnasts are familiar with the use of nutritional aids. The most frequently used supplements are vitamins, proteins and calcium. A questionnaire study among Greek elite gymnasts showed that 58% have taken nutritional supplements, following their doctors' recommendations (Zaggelidis *et al.*, 2005). As in other sports, a significant number of gymnasts are not aware of the prohibited substances list, nor of the possible detrimental effects of excessive supplementation.

A four-month period of intense training and supplementation with soy protein in several members of the Romanian Olympic female team has demonstrated an increase in lean body mass and serum levels of prolactin and T4, and also a decrease in serum alkaline phosphatases (Stroescu *et al.*, 2001). However, a 12-month randomized control trial comparing the effect of 500-mg calcium supplementation between gymnasts and a group of similar ages did not show any significant difference in volumetric bone mineral density (BMD) for the radius, tibia, spine and whole body (Ward *et al.*, 2007). The authors concluded that there were no beneficial effect in gymnasts of additional calcium beyond the recommended levels (555–800 mg/day for 8- to 11-year-olds). They also suggested that the lack of significant difference in volumetric BMD could be related to the fact that gymnasts' bones adapt under the effect of repetitive loading and therefore, do not benefit from additional calcium supplementation.

To conclude, most authors confirmed several nutrient deficiencies; however, investigations on supplementation were not conclusive. In fact, most of the studies did not check if there were any deficiencies at baseline and this makes interpretation quite difficult. While balanced nutrition might be sufficient for some gymnasts, common sense would suggest that supplements are required for those who have some deficiencies. Nonetheless, these supplements should be prescribed by a sport dietician in order to avoid any risks of taking banned substances. The remaining issue, and the most important thing to consider and monitor, is: do they eat enough and what do they eat?

## 6.3 Effects of high volume and intensity of training on body composition, hormonal regulation, growth and sexual development

Several health issues related to gymnastics (artistic and rhythmic) have been at the centre of ongoing debates for the last 15 years. Among these issues: body composition and its relation to a high volume of training, combined with inadequate diet; and also hormonal regulation, growth and sexual development. Online search engines showed up to 229 articles based on the following key words: *gymnastics*, *growth*, *energy intake*, *amenorrhoea* and *bone mineral density*. This section highlights some of the main health problems that gymnasts encounter at the highest level of performance.

### *6.3.1 Body composition*

It has become evident that gymnasts have a very low body fat percentage and body mass index (BMI) when compared to age- and gender-matched non-gymnast groups; this is particularly noticeable in females (rhythmic and artistic) (Cassell *et al.*, 1996; Claessens *et al.*, 1992; Courteix *et al.*, 1999; Malina, 1994; Soric *et al.*, 2008). Of the reviewed articles, 98% confirmed a cross-sectional decrease in body composition parameters, in particular weight and height, while 43% confirmed a longitudinal effect. Table 6.2 shows body mass index and body fat percentages reported in male and female gymnasts. It has also been demonstrated that these decreased body composition variables are sustained over the years. Benardot & Czerwinski (1991) reported a cross-sectional BMI range of 12.9 to 20.8 kg/m$^2$ in elite junior gymnasts, aged between 7 and 10 years old and 14.6 to 20 kg/m$^2$ for those between 11 and 14 years old. The percentage of body fat reported in the same group was as follows: 5.1 to 16.7% between 7 and 10 years, which does not change between 11 and 14 years and remains around 6 to 15.1%.

### *6.3.2 Bone development and mineral density*

It has been shown that low energy intake associated with high energy expenditure might be considered as two potential complicating factors which may affect bone development and later osteoporosis (Benardot *et al.*, 1989; Drinkwater *et al.*, 1984). Around 80% of the studies

**TABLE 6.2** Body mass index and fat percentage in male and female gymnasts

| *Females* | *Courteix et al.* (2007)<br>13.4 yrs<br>n = 36<br>RG | *Douda et al.* (2006)<br>13 yrs<br>n = 39<br>RG | *Jemni et al.* (2006)<br>20.6 yrs<br>n = 21<br>AG | *Georgopoulos et al.* (2004)<br>16 yrs<br>n = 169<br>AG | *Markou et al.* (2004)<br>16 yrs<br>n = 120<br>AG | *Muñoz et al.* (2004)<br>16.2 yrs<br>n = 9<br>RG | *Bale et al.* (1996)<br>13.3 yrs<br>n = 20<br>AG |
|---|---|---|---|---|---|---|---|
| BMI (kg/m²) | 16.5 | 15.9 | | 19.0 | 18.6 | 18.6 | 18.2 |
| % BF | 14.4 | 13.9 | | 19.5 | 18.4 | – | 10.9 |
| *Males* | | | | 17 yrs<br>n=93<br>AG | 18 yrs<br>n=68<br>AG | | |
| BMI (kg/m²) | – | | 23.2 | 21.5 | 21.1 | – | – |
| % BF | – | | 9.7 | 10.6 | 10.3 | – | – |

AG: artistic gymnastics; RG: rhythmic gymnastics.

confirmed the skeletal maturation delay in artistic gymnasts, whereas, this agreement reaches almost 98% in rhythmic gymnastics. Theodoropoulou *et al.* (2005) undertook one of the wider-scale investigations by assessing 400 rhythmic gymnasts and 400 female artistic gymnasts from 32 countries, all of them world-class females aged between 11 and 23 years. The authors confirmed a delayed bone development of 1.3 years in rhythmic gymnasts and of 2.2 years in artistic gymnasts. Similar findings were obtained by Muñoz *et al.* (2004): around a two year delay in rhythmic gymnasts compared to controls. However, it seems that male gymnasts are less affected; Markou *et al.* (2004) proved that bone age was delayed by two years in 169 females, compared to only one year in 93 males aged between 13 to 23 years, all at elite levels. Similarly, Jemni *et al.* (2000) have also shown a six-month developmental delay in post-pubertal high-level French male gymnasts (18.5 years old).

It appears that artistic gymnasts endure more intense mechanical loads for the upper and lower limbs and trunk because of the higher acrobatic skills, including mounts and dismounts. Authors have confirmed that gymnastic exercises are associated with an increase in bone mineral content (BMC) and density (BMD), particularly in the most exposed joints (wrists and ankles) and the lumbar spine (Courteix *et al.*, 2007). It was suggested that the higher levels of osteocalcin might be the main contributor to higher BMD. Furthermore, the authors suggested that preserved bone health might be a counterbalanced effect of hormonally disturbed athletes. In addition, it appears that gymnastics has a long-lasting beneficial effect on bones. It has been shown that total or partial BMD increases after retirement in active and/or inactive gymnasts (Dowthwaite & Scerpella, 2009). Ducher *et al.* (2009) provided evidence of an enhanced mineral content of the radius and ulna in retired artistic gymnasts (18–36 years). These findings confirm previous studies: Markou *et al.* (2004) proved that long-term high-intensity exercise is negatively correlated with BMD, whereas it is positively correlated with their chronological age. In their review of the literature, Dowthwaite & Scerpella (2009) recommended further research to elucidate skeletal loading dose–response curves and the sex- and maturity-dependence of skeletal adaptation, which varies across anatomical sites and tissue compositions.

As a final point, it can be concluded that BMD excess is exercise-related.

### 6.3.3 Hormonal regulation, growth and sexual development

Markou *et al.* (2004) have reported around 32% and 12% of female amenorrhoeic rhythmic gymnasts (absence of menstruation by the age of 16 years) in Canada and in Greece, respectively. Meanwhile the percentage of the oligoamenorrhoeic rhythmic gymnasts (menstrual cycle duration greater than 36 days) in both countries was more dramatic (around 65%). Similar findings were also confirmed in Spanish rhythmic gymnasts with 45% oligoamenorrhoeic (Muñoz *et al.*, 2004). Pubertal delays have been confirmed in both artistic and rhythmic gymnastics, with a more pronounced effect in artistic gymnastics due to higher stress imposed by training intensity and the higher number of competitions. Muñoz *et al.* (2004) revealed that rhythmic gymnasts reached their menarche at 15 ± 0.9 years compared to controls (12.5 ± 1 years). Similarly, Courteix *et al.* (2007) found a two-year delay in French rhythmic gymnasts, whereas Theodoropoulou *et al.* (2005) found not only that puberty was delayed in 400 rhythmic and 400 artistic gymnasts, but also that 17% and 20% of them respectively had no menarche also, by comparison with their mothers' menarche and to their untrained sisters. Likewise, Sands *et al.* (2002) confirmed the differences in menarche age and menstruation irregularities between high-level American gymnasts and their mothers via a questionnaire study. These findings, in reality, demonstrate that growth and development issues are not related to any genetic variables, but are actually the consequence of training stresses.

Several authors had a particular interest in investigating leptin in gymnasts. Leptin is in fact, an ob-gene protein secreted by fat cells; it has a role in the regulation of body weight and in the stimulation of the reproductive axis (through fat). Evidence shows that mechanisms leading to a deregulation of the reproductive axis in patients with anorexia nervosa are comparable with those leading to delayed puberty in elite female gymnasts with very low body fat and low diet. This may lead to the 'Female Triad', which is a combination of eating disorders, amenorrhoea and osteoporosis.

Some 65% of the studies confirmed a decreased level of leptin in both rhythmic and artistic gymnasts (Courteix *et al.*, 2007; Jemni *et al.*, 2008; Markou *et al.*, 2004; Muñoz *et al.*, 2004; Weimann, 2002; Weimann *et al.*, 1999). Notably, it has been shown that leptin levels of gymnasts with particularly low fat stores, are less than those measured in a group of patients with anorexia nervosa (1.2 ± 0.8 μg/l versus 2.9 ± 2.7 μg/l respectively) (Matejek *et al.*, 1999). This hypoleptinaemia was attributed to insufficient caloric intake. Weimann *et al.* (2000) have suggested that hypoleptinaemia, in turn, causes delayed puberty and growth. In fact, several authors have confirmed decreased oestradiol and oestrogen levels in pubertal gymnasts and there was no significant rise occurring with normal sexual maturation, whereas, IGF-1, IGF BP-3, TSH, T3 and T4 showed normal age-dependent serum levels (Courteix *et al.*, 2007; Weimann, 2002; Weimann *et al.*, 1999).

Leptin and oestrogen production by fat tissue plays a crucial role in triggering menarche, reflecting a natural adaptation of the body to high-energy demands (Moschos *et al.*, 2002). Female fat tissue is a significant source of oestrogens (by the conversion of androgens to oestrogens). A decreased conversion of androgens to oestrogens because of decreased fat tissue in athletes may contribute to delayed breast development and menarche (Perel & Killinger, 1979; Schindler *et al.*, 1972). However, recent experiments have suggested that menstrual disorders are part of a metabolic response to an energy-deficit relationship: the stress of high-energy output ± inappropriate low-energy intake ± energy cost ± energy availability (Loucks & Redman, 2004).

## 6.4 Conclusion

- There is considerable pressure on gymnasts (especially females, rhythmic and artistic) to maintain leanness.
- There is evidence of eating disorders which have serious health implications.
- In both genders, total energy consumption and nutritional intake are insufficient in spite of high energy expenditure although to a lesser extent in male gymnasts, who have a more balanced diet.
- There is evidence that female gymnasts suffer from several mineral deficiencies – in particular, deficiencies in calcium, thiamine and riboflavin, iron, fibre, vitamin E and vitamin B6.
- Significant percentages of elite gymnasts are familiar with the use of nutritional aids. The most frequently used supplements are vitamins, proteins and calcium.
- Only a few studies have investigated the effects of supplementation and the results are therefore inconclusive. However, protein supplementation has shown some effects on body composition, whereas calcium did not show any benefits to bone health.
- Gymnastics is a significant stimulus to maintain a very low body fat percentage and a very low BMI; this is particularly noticeable in female rhythmic and artistic gymnasts.
- There is strong agreement on skeletal maturation delay, which affects artistic gymnasts more than rhythmic ones, because of the increased stress of training and competitions. Gymnastics increases bone mineral content and density, particularly in the most exposed joints and the lumbar spine. There is accumulating evidence that benefits persist into adulthood and retirement, both in those who remain active gymnasts and those who do not.
- Pubertal delays have been confirmed in both artistic and rhythmic gymnastics, with a more pronounced effect in artistic gymnastics.
- The literature shows a very high number of female amenorrhoeic gymnasts and alarming numbers of oligoamenorrhoeic ones.
- There is insufficient evidence to conclude that amenorrhoeic gymnasts do not experience premature osteoporosis.
- There is a confirmed decrease of leptin levels in both rhythmic and artistic gymnasts with particularly low fat stores, with some values less than those measured in a group of patients with anorexia nervosa.
- It is suggested that hypoleptinaemia causes delayed puberty and growth because of decreased oestradiol and oestrogen, indirectly caused by the reduced fat store.
- There is evidence that growth and development issues are induced not by genetics, but by exercise.

# 7

# INVESTIGATIONS IN RHYTHMIC GYMNASTICS

*Maria Gateva*

## 7.1 Introduction and objectives

Rhythmic gymnastics (RG) is a sport that combines elements of ballet, gymnastics, dance, and apparatus manipulations. Rhythmic gymnasts compete in two different categories: individual and group competitions (five gymnasts compose the group). Gymnasts perform routines with one of the following five types of apparatus at once: rope, hoop, ball, clubs, and ribbon. The ranking is made by a panel of judges, who give marks for each of the following aspects: technique, artistry and execution.

The routine is a creation of 'music story-based' choreography, including some compulsory body movements combined with technical elements using the apparatus. The duration of the routines is 1 min 15 sec to 1 min 30 sec for the individual exercises and 2 min 15 sec to 2 min 30 sec for the group exercises.

The international gymnastics federation (FIG) recognized rhythmic gymnastics as a new discipline in 1961. The first world individual championships were held in Budapest, Hungary, in 1963. The first Olympic Games to feature RG as a discipline of its own were in 1984 in Los Angeles, California, for individuals and in 1996 in Atlanta, Georgia, for groups. World Championships and Grand-Prix tournaments are held every year.

The aim of this chapter is to provide a current overview of the modern rhythmic gymnasts' physical fitness and physiological characteristics. The chapter also supplies an up-to-date battery of tests to assess the most relevant physical aptitudes.

## 7.2 Importance of physical preparation and assessment in rhythmic gymnastics

Physical preparation is one of the most important pillars enabling peak performance and, ultimately, the winning of medals, in any sport. Rhythmic gymnastics has undergone immense development in the last few decades, owing to the ever increasing technical skills required through revision of the Code of Points. These rules stress the importance of high-difficulty activities. High difficulties require suitable physical fitness. Rhythmic gymnasts are expected to have maximal flexibility (even beyond the normal range of motion), coordination and balance. In addition, optimal strength and power are also crucial variables of the performance.

Section 6 provides evidence of obsessive behaviour towards maintaining leanness, particularly among rhythmic gymnasts. These behaviours lead to eating disorders, insufficient energy intake versus high energy expenditure and several mineral deficiencies. It has been proved that these conditions lead also to serious health implications, such as amenorrhoea or delayed menarche (Georgopoulos *et al.*, 2002; Jankauskienė & Kardelis, 2005; Lindholm *et al.*, 1995; Markou *et al.*, 2004). Moreover, it has been shown that negative energy balance is directly related to an increased incidence of stress fractures in rhythmic gymnastics (Benardot, 1999). The same author confirmed that not only do many elite rhythmic gymnasts have one injury or more per year but also 86% of them complain of low back pain, with more injuries occurring as major events approach.

Conditions such as exaggerated lordosis of the spine, stress fractures on one or both sides of the vertebrae (spondylolysis), spondylolisthesis (with or without obvious symptoms), Sheuermann's disease and disc degeneration are common among rhythmic gymnasts. The most common symptoms of these conditions are pain that spreads across the lower back, which may feel like a muscle strain, or spasms that stiffen the back and tighten the hamstring. These symptoms systematically lead to changes in posture and gait. Ignoring these symptoms and continuing the activity without regular monitoring may complicate the conditions and result in having to cease all activities in order to undergo spinal surgery.

Returning to training and competition after surgery may take a long time. During this time, rehabilitation should focus on strength improvement of the trunk and lumbar spine (Kruse & Lemmen 2009). Also, strengthening and stretching exercises for the back and the abdominal muscles help prevent future recurrences of the pain.

Although medical workers underline the importance of prevention, the optimizing of training loads is still the centre of long-lasting debates between scientists and coaches in order to avoid these health issues. From this point of view, on the basis of many years of competition at a high level, the value of physical preparation is generally undervalued by coaches. However, appropriate physical preparation with relevant periodic monitoring should be as important as technical preparation. Specialists have to pay enough attention to this problem to minimize injuries. In addition, progressive loading combined with appropriate regeneration, taking into consideration the readiness of the gymnast (see Chapter 4), should also be rigorously applied. Monitoring should also become a regular 'sporting culture' in rhythmic gymnastics. The following section provides, indeed, some monitoring tips.

## 7.3 Physical fitness assessment in rhythmic gymnastics

This section presents a series of physical fitness tests which might be applied to different age and level groups, for different purposes (detection, selection, or cross-sectional or longitudinal evaluations). These tests are specific to several clusters of physical qualities, which are:

- speed
- agility and power
- specific muscular endurance
- flexibility
- coordination
- balance
- musicality and sense of rhythm.

Figure 7.1 shows these tests; they were originally created by Vankov in 1983 (Vankov, 1982, 1983), and approved and adopted by the Bulgarian Rhythmic Gymnastics Federation. More recently, they have been modified and updated by Gateva in 2005 (Gateva & Andonov, 2005). Regular applications of these tests may help in monitoring and optimizing training loads.

### 7.3.1 Description of the tests

Note: the tests listed below and illustrated in Figure 7.1 are not set in the order of performance. We recommend starting with the speed and agility tests followed by one of the muscular endurance tests, then by one of the balance, flexibility or rhythm and musicality tests to alternate again with one of the muscular endurance tests. A total recovery time should be given to the gymnast before attempting the muscular endurance tests. Alternatively, it is recommended the performance of the battery of tests on separate days.

- No. 1 – Shuttle sprint (3 × 10m) (in seconds)
- No. 2 – Alternative right- and left-foot jumps with a rhythmic gymnastics rope: 20 jumps are performed at maximum speed. The assessment begins with the first jump going through the rope (in seconds). Only one trial is allowed.
- No. 3 – Sit-ups (dynamic force of the abdominal muscles): 15 repetitions at maximum speed (in seconds).
- No. 4 – Maximal vertical jump (in cm).
- No. 5 – Maximal repeated jumps: from lying in the back position with full body extension, stand up and vertical jump with arms swing then back to the initial position, repeat until exhaustion. The number of repetitions is counted, and the tempo is set with a steady musical rhythm.
- No. 6 – Arm push-ups until exhaustion. The number of repetitions is counted, and the tempo is set with a steady musical rhythm.
- No. 7 – Repeated side jumps over the rope until exhaustion. The height of the rope is half of the height achieved at the maximal vertical jump (Test 4). The number of repetitions is counted, and the tempo is set with a steady musical rhythm.
- No. 8 – Flexibility of the spine: The gymnast is bent backwards with the help of a partner holding her hands. Legs are held together. Once the bent position is stable, the vertical distance from the upper limit of the shoulder-blades to the upper limit of the iliac bone is measured in cm (which may require marks on the body before the test).
- No. 9 – Dynamic flexibility of the coxo-femoral joints: Forward and sideways dynamic kicks with the right and left legs. The range of motion of the legs is measured in degrees using a chart (see Figure 7.1). The best of the three trials is recorded.
- No. 10 – Walking on toes with closed eyes along straight line 5 m long. Any deviation from the straight line is measured in cm and rounded up to the nearest 0.5 cm. The final result is the sum of all of the deviations.
- No. 11 – Walking on the toes along a straight line 5 m long, following 180-degrees spinning with the body bent at 90 degrees. The deviation from the straight line is measured in cm and rounded up to the nearest 0.5 cm. The final result is the sum of all the deviations.
- No. 12 – Held balance in passé position with the arms upward. Three trials are performed. The average time is recorded exactly, to the nearest 0.1 second. The time for the left and right leg is measured separately.
- No. 13 – Coordination test: A path 14 m long with a width of 60 cm is outlined on the carpet. The gymnast starts with one step and performs a series of three 360-degree pivots on both feet with maximum speed and accuracy, followed by three forward rolls, a forward walkover and a backward walkover. Then she has to perform three pivots, followed by three forward rolls in the opposite direction. Once finished, she has to immediately hold a 3-sec balance in the passé position. The whole duration of the routine is measured in seconds (without the time for the held balance). If the gymnast fails to perform the

balance, a penalty time is given. The penalty is equal to the time during the three-second balance in which the gymnast failed to keep balance. A penalty is also given for every touchdown out of the restrictive area. The distance is measured from the outer line to the touchdown mark in cm and transformed into seconds (1 sec = 1 cm). The duration of the routine is then added to the penalty time to provide the final test score. (This test requires two investigators.)

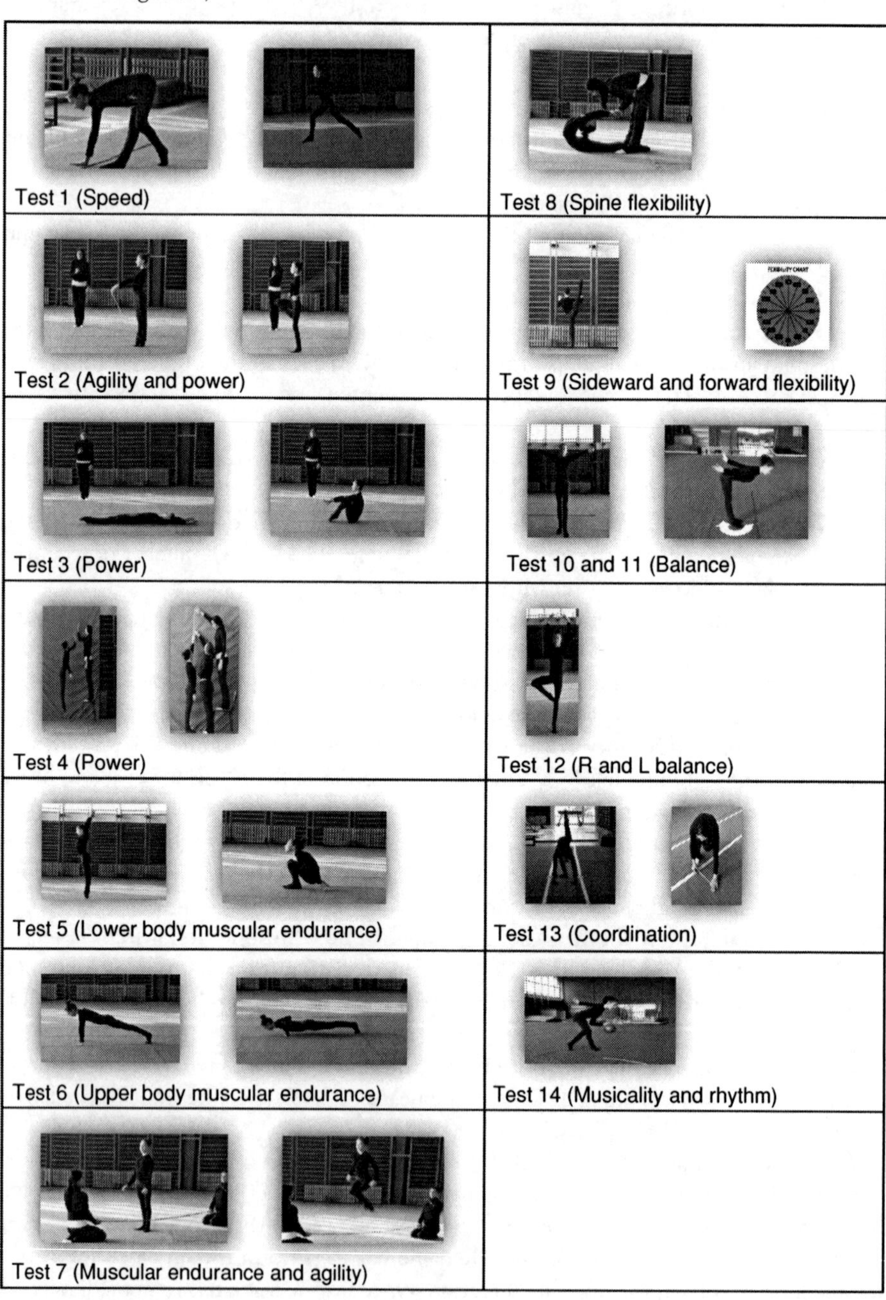


**FIGURE 7.1** Physical fitness tests for rhythmic gymnasts. (Gateva and Andonov, 2005)

- No. 14 – Musicality and sense of rhythm test: The musicality and sense of rhythm are assessed with two displays, to music at different tempos. One is a regular tempo, the other is non-regular. The gymnast has to follow the rhythm in three different ways for each type of tempo: clapping her hands, bouncing a ball on the carpet and jumping over a rope with alternate feet. Expert judges give marks from 2 to 6 for each trial. The final mark is the average of all six marks.

## 7.4 Training loads in high-level rhythmic gymnastics

The complicated structure of rhythmic gymnastics sessions makes quantifying training loads a laborious task. The effect of an extra 40 minutes or 60 minutes of specific physical preparation during the competitive period and the preparation period, respectively, added to the normal training session,was compared between an experimental and a control group of rhythmic gymnasts. The experimentation lasted two years and the above physical fitness tests were applied regularly to readjust training loads. Table 7.1 presents the baseline measurements and measurements made after two years.

**TABLE 7.1** Physical fitness assessment of rhythmic gymnasts in average and (SD). Two years longitudinal investigation (Gateva and Andonov, 2005)

| *Tests* | *Experimental group* n= 20, 9–11 *yrs* | | *Control group* n = 20, 9–11 *yrs* | |
|---|---|---|---|---|
| | *Baseline* | *Post* | *Baseline* | *Post* |
| No. 1: speed (sec) | 9.5 (0.6) | 8.6 (0.4)*^ | 9.8 (0.6) | 9.4 (0.6)^ |
| No. 2: agility 8, power (sec) | 9.5 (1.0) | 6.3 (0.4)^ | 9.8 (0.6) | 6.4 (0.7)^ |
| No. 3: power (sec) | 16.9 (1.4)* | 13.7 (0.9)*^ | 18.8 (3.1) | 15.4 (1.7)^ |
| No. 4: power (cm) | 30.3 (5.9) | 37.2 (3.8)^ | 31.2 (4.0) | 35.7 (4.8)^ |
| No. 5: LB muscular endurance (rib) | 80.9 (58.2)* | 272.6 (194)*^ | 42.7 (27.0) | 54.8 (35.4) |
| No. 6: UB muscular endurance (nb) | 14.3 (8.3) | 21.4 (9.2)*^ | 9.6 (7.5) | 16.6 (8.4)^ |
| No. 7: muscular endurance 8, agility (nb) | 59.5 (14.4) | 91.3 (8.8)*^ | 58.5 (20.4) | 58.8 (19.2) |
| No. 8: spine flexibility (cm) | 13.6 (7.2) | 10.4 (5.1) | 13.8 (4.3) | 11.1 (3.2) |
| No. 9: flexibility R leg (forward) (degrees) | 162.8 (14.5) | 176.5 (5.4)^ | 168.0 (7.9) | 176.4 (9.4)^ |
| No. 9: flexibility L leg (degrees) (forward) | 152.8 (14.3)* | 163.1 (9.0)^ | 161.5 (9.3) | 163.0 (10.2) |
| No. 9: flexibility R leg (sideward) (degrees) | 162.1 (16.1)* | 170.0 (9.0)^ | 171.9 (8.0) | 177.1 (11.2) |
| No. 9: flexibility L leg (sideward) (degrees) | 151.4 (17.2) | 162.8 (10.0)^ | 159.8 (10.1) | 163.4 (10.3) |
| No. 10: balance (cm) | 40.5 (18.4)* | 16.0 (8.5)^ | 28.4 (13.6) | 15.5 (12.3)^ |
| No. 11: balance (cm) | 29.1 (15.7) | 9.6 (5.8)^ | 27.4 (12.0) | 10.6 (9.4)^ |
| No. 12: R leg balance (cm) | 4.3 (4.8) | 7.9 (6.5) | 7.7 (10.5) | 9.9 (6.8) |
| No. 12: L leg balance (cm) | 2.33 (1.72)* | 8.83 (9.03)^ | 5.81 (6.04) | 14.1 (15.8)^ |
| No. 13: coordination (sec) | 28.6 (4.0) | 23.3 (2.0)^ | 28.7 (8.3) | 22.9 (8.0)^ |
| No. 14: musicality and rhythm (marks) | 4.9 (0.7) | 5.8 (0.3)^ | 5.2 (0.6) | 5.5 (0.5)^ |

*: Significant difference between the groups ($p<0.05$).

^: Significant difference from baseline to the end ($p<0.05$).

As expected, the physical abilities of the experimental group significantly improved compared to the controls. Particular improvements were shown in the muscular endurance tests, flexibility and coordination.

The performance of the experimental group improved tremendously, and they took the bronze medal in the national championship, compared to a lower position in the previous years. In addition, all the experimental gymnasts were ranked between the 8th and the 19th position.

In addition to the specific physical preparation, the experimental group performed the following loads in each training season:

- the total training volume per season: 263 days
- 393 training sessions per season
- 1,739 full-length routines
- 2,327 parts of routines
- 83,601 exercises – that is, 318 exercises per session.

## 7.5 Biochemical and energetic investigations in high-level rhythmic gymnasts

To assess the energetic pathways applied while performing rhythmic gymnastics exercises, serum levels of creatine phosphokinase (CPK), isoenzyme of creatine phosphokinase (CPK-MB, also called CPK-2) and lactate dehydrogenase (LDH) have been measured in 10 high-level rhythmic gymnasts, members of the first and second groups of the Bulgarian national team during the preparation period. Measurements took place at three different times in a normal training day: 8 h before the training sessions; 12 h immediately after the first training session; 14 h before the second training session. Analysis was performed using a Humalyzer 2000 (Germany) according to Hoffman La Roche procedure.

The results of this investigation are shown in Table 7.2. However, to understand the meaning of these results, a brief explanation of CPK, CPK-MB and LDH is necessary.

*Creatine phosphokinase (CPK)* is an enzyme found in skeletal muscle. The normal range is 40 to 210 units per litre in normal adults, but values may vary according to the type of analysis and the laboratory procedures (Goh & Fock, 1985). Regular exercise increases the resting level of serum CPK, and therefore it is higher in trained individuals than in untrained ones (Fitts, 1995). High-intensity exercise amplifies the amounts of CPK in the extracellular fluid and plasma. This might be involved in fostering muscle cell permeability and/or the muscle degeneration and necrosis that occur during exercise (Clarkson *et al.*, 2006; Kaman *et al.*, 1977). In addition, muscular soreness has been reported to elevate CPK levels (Fitts, 1995). The type and duration of exercise affect the elevation of serum CPK. Swimming, for example, causes less CPK elevation than weight-bearing events (Argeitaki *et al.*, 2009). Peaks in serum CPK post-exercise have been reported to occur immediately after exercise and even up to the following six days (Clarkson *et al.*, 2006; La Porta *et al.*, 1978; Oliver *et al.*, 1978; Riley *et al.*, 1975).

*The MB isoenzyme of creatine phosphokinase (CPK-MB)* is a specific indicator of myocardial injury. Anderson (2007) has demonstrated that CPK-MB levels rise 3–6 hours after a heart attack. The peak level may occur after 12 to 24 hours and returns to normal 12 to 48 hours after tissue death. In addition, Anderson (2007) and Barohn (2007) have shown that an increased CPK-MB levels may also be due to:

- electrical injuries
- heart defibrillation
- heart injury

- inflammation of the heart muscle, usually due to a virus (myocarditis)
- open-heart surgery
- vigorous and prolonged exercise or immobilization.

*Lactate dehydrogenase (LDH)* is an enzyme of the glycolytic metabolism that catalyses the reciprocal reaction of lactate and pyruvate with interconversion of $NAD^+$ and NADH. Pyruvate is converted to lactate when oxygen is in short supply. The reverse reaction is performed during the Cori cycle in the liver (Butt *et al.*, 2002). Some of the body organs are relatively rich in LDH, such as the heart, kidney, liver, brain, lungs and muscles. A typical range of the LDH for untrained persons is 105–333 u/l (units per litre) (Abraham *et al.*, 2006). It has been shown that strenuous exercise may raise the level of total LDH, in particular its isoenzymes LDH-1, LDH-2 and LDH-5 (Flores, 2001). The LDH test is also commonly used to measure tissue damage.

Abraham *et al.* (2006) and Schwartz (2007) have shown that higher-than-normal levels may indicate:

- blood flow deficiency (ischaemia)
- cerebrovascular accident (such as a stroke)
- heart attack
- haemolytic anaemia
- liver disease (for example, hepatitis)
- low blood pressure
- muscle injury
- muscular dystrophy.

The literature provides LDH values in different sports which vary from 262.5 u/l for road cyclists up to 371.8 u/l for long-distance runners (Argeitaki *et al.*, 2009).

*Lactic acid (BL)* is a natural bio-product of anaerobic glycolysis. It is produced and accumulated rapidly when the intensity of the effort is quite high and in particular, when oxygen is in short supply. The acidic environment impairs muscle cell contraction and provokes an increasing level of fatigue, which lead to the reduction of physical performance and exhaustion. When oxygen is available, lactic acid is reconverted to pyruvate (Butt *et al.*, 2002).

## 7.6 Analysis of the main results

It should be noted that this is the first investigation that provides such biochemical variables in gymnasts (all disciplines included) and therefore, any comparisons are quite impossible.

**TABLE 7.2** Biochemical variables in high-level rhythmic gymnasts (averages and SDs in units per litre)

| | *8:00 h Baseline before training* | *12:00 h Immediately after the first training session* | *14:00 h 2 hours after the first training session* |
|---|---|---|---|
| LDH (units/l) | 330.5 ± 66.6 | 388.3 ± 75.1 | 445.3 ± 93.6* |
| CPK (units/l) | 153.0 ± 222.1 | 208.7 ± 266.3 | 297.6 ± 345.8* |
| CPK-MB (units/l) | 8.0 ± 5.7 | 12.2 ± 5.4 | 10.9 ± 5.4 |
| BL (mmol/l) | 1.4 ± 0.4 | 1.3 ± 1.0 | 2.2 ± 1.5* |

* Significantly different ($p < 0.05$).

As shown in Table 7.2, there are increases of the glycolytic markers (i.e., LDH and CPK) from the baseline to post-training with even more significant increases two hours after the session. These values may indeed further increase after the second training session. This finding suggests that the two hours' break is insufficient to restore the energetic substrates, particularly when we recall the insufficient energy intake highlighted in the above sections. The findings suggest also that the three-hour training session was not quite intense enough and not a sufficiently significant stimulus to engage the anaerobic system as the main energy provider. Blood lactate was indeed very low across the three samples. As a consequence, LDH increase does not reflect any muscle damage. The training intensity was actually low because of the nature of the preparation period. Gymnasts were essentially learning new skills and mini routines for the new season. However, LDH increase might reflect the conversion of the little amount of lactate back to pyruvate.

It is also important to notice the huge standard deviations of the CPK values in the three analyses. Two of the reasons which may explain the high standard deviations are that some gymnasts, particularly the second team (substitutes), had reduced training programmes compared to the first team. More attention was indeed given to the first team members, with a rigorous monitoring of the training loads. In addition, some gymnasts were recovering from injuries and their training loads were therefore also decreased.

The results show also that CPK-MB values do not show any significant changes. This suggests appropriate adaptations of the myocardium to the applied training loads.

## 7.7 Conclusion

- Physical preparation in rhythmic gymnastics should take a more substantial place in training and preparation, in view of the increasing difficulties of the Code of Points, and also to help in reducing the risk of injuries.
- Physical fitness assessment should be regularly applied and should include the seven highlighted physical fitness clusters: speed; agility and power; specific muscular endurance; flexibility; coordination; balance; and musicality and sense of rhythm.
- High-level rhythmic gymnastics loads are quite substantial high.
- Biochemical markers show that anaerobic metabolism is not heavily involved in supplying energy.
- LDH enzyme increases after training because of the increase in lactate, and therefore, enhances the recovery process by converting lactate back to pyruvate.
- CPK-MB coenzyme values suggest good adaptation of the myocardium to the training loads applied in high-level rhythmic gymnasts.

## Acknowledgement

A word of thanks is due to all the gymnasts and their coaches who participated in the above studies and to Dr Petar Atanasov, whose help was crucial for the biochemistry analysis.

# PART I REVIEW QUESTIONS

Q1: Describe gymnasts' maximal oxygen uptake and comment on its importance to gymnastic performance.

Q2: Analyse the evolution of the $VO_2$ max of gymnasts throughout the last few decades and relate it to the evolution of the sport.

Q3: Do you think it is important to enhance gymnasts' $VO_2$ max and why?

Q4: Explain the physiological and training concepts affecting lactic threshold delay in gymnasts.

Q5: Describe a few methods and techniques used to estimate and/or to measure the energy cost of male and female gymnastic exercises/routines.

Q6: Using relevant data and illustrations, discuss the energetic cost of gymnastics exercises *(essay-type question)*.

Q7: Analyse the peak power output of gymnasts as assessed by standardized tests.

Q8: Analyse the metabolic responses to gymnastics routines and exercise.

Q9: Compare the aerobic and anaerobic metabolisms of gymnasts *(very long essay-type question)*.

Q10: Analyse the respiratory system in gymnasts.

Q11: Analyse the cardiovascular adaptations to gymnastic exercises.

Q12: Analyse heart-rate responses during gymnastic exercises.

Q13: Analyse the strength, speed, power, flexibility and muscular endurance of the gymnasts.

Q14: Analyse each of the following training principles in relation to gymnastics: *(long essay-type question or average- length answer if taken individually)*.

- specificity
- readiness

- individualization
- variation
- diminishing returns
- regeneration
- overload and progression
- periodization.

Q15: Give examples of specific tests used to assess jumping abilities in gymnastics.

Q16: Give examples of specific tests used to assess muscular endurance in gymnastics.

Q17: Give examples of specific tests used to assess agility, speed, strength and power in gymnastics.

Q18: Give examples of specific tests used to assess flexibility in gymnastics.

Q19: Give example of specific technical tests used in gymnastics.

Q20: Discuss the relation between nutritional intake and energy expenditure in male and female gymnasts *(essay-type question)*.

Q21: Discuss ergogenic (performance-boosting) supplementation in gymnastics.

Q22: Discuss the effect of high volume and intensity of training on the gymnasts' body composition, hormonal regulation, growth and sexual development *(long essay type question or average length answer if taken individually)*

Q23: Analyse the outcomes of assessing the following biochemical markers in high-level rhythmic gymnasts: LDH, CPK and CPK-MB.

# PART II

# Biomechanics for gymnastics

## Introduction and objectives

*William A. Sands*

### What is biomechanics?

Biomechanics is a part of physics that studies the mechanical or physical principles as they apply to the movement of living things. Biomechanics serves gymnastics by applying the principles and techniques of physics and mechanics to the movement of a gymnast and the apparatus.

### Why biomechanics?

Biomechanics is an important part of all sports, not just gymnastics. Biomechanics relies on physics, which in turn relies on the character of physical laws. Physical laws are important for a variety of reasons and rely heavily on mathematics for unambiguous expression. One of the most important concepts behind physical laws is that once a law is determined, then further laws can be derived from the original law. For example, if we have a stopwatch and a tape measure and we know that a gymnast running for vaulting covered a distance of 10 metres in 2 seconds, then the average speed over that distance was 5 metres per second (10m/2s=5m/s). If we have a radar gun instead and we know that the gymnast travelled 5 metres per second for 2 seconds then we can determine that the gymnast travelled 10 metres (5m/s x 2s = 10m).

This may be overly simplistic, but the beauty of the mathematics in physical laws is that we can use these concepts repeatedly, with modifications, to learn new things.

### Units of biomechanics

When compared to the other scientific areas of sport and exercise, biomechanics has relatively few independent concepts and is thereby much easier to grasp. Table II.1 shows most of the units of mechanics that are used in gymnastics biomechanics.

Note there are only four fundamental units, and that all derived units are simply combinations of these fundamental units. This simplicity comprises some of the elegance of physical laws and the application of physical laws to a sport like gymnastics.

In addition to having a relatively small number of independent concepts, compared with, say, physiology or psychology, the biomechanics of gymnastics can be further subdivided based

**TABLE II.1** Units of mechanics

| *Fundamental units* | *Derived units* | *Units with special names* |
|---|---|---|
| Time (s) | Velocity (m/s) | Force (newton, N = kg.m/s²) |
| Displacement (m) | Acceleration (m/s²) | Pressure and stress (pascal, P = N/m²) |
| Mass (kg) | Momentum (kg (m/s)) | Energy and work (joule, J = Nm) |
| Temperature (Celsius, °C) (Kelvin° – 273.15) | Torque (Nm⊥) | Power (watt, W = J/s) |
| | Moment of force (Nm⊥) | Angle (radians or degrees) |
| | Density (kg/m³) | |
| | Area (m²) | |

The ⊥ symbol indicates that the distance is measured perpendicularly from the line of application of force rather than collinearly, as in energy and work.

on the type of motion involved and whether we are interested in just describing the motion or in knowing the origins of the motion.

## What is the role of biomechanics in gymnastics?

Biomechanics is the study of technique. Techniques are commonly studied and evolve over time. When addressing the technical errors involved in gymnastics, often the instructor sees the technique fault but is forced to rely on trial-and-error teaching in order to fix the fault. Moreover, the actual fault is often performed long before the technique problem becomes visible to the naked eye. For example, when a gymnast is airborne in a somersault and shows a fault in technique, that fault is almost always traceable to the takeoff or skills preceding the takeoff. Biomechanics is the science that can discover the source of the technical fault rather than simply identify the obvious.

## Conclusion

We further reduce the number of concepts in biomechanics by dividing our thinking into four areas: linear kinematics, angular kinematics, linear kinetics and angular kinetics. Kinematics refers to a description of motion via the related concepts of position, displacement, velocity and acceleration. Kinematics can furnish a nearly complete description of the motion with just these concepts and most coaches and teachers are experts at kinematics because this is the most visible part of performance.

Kinematics can be 'seen' by the athlete, teacher or coach, and thereby forms the bulk of the information used for the qualitative assessment of motion. Kinetics adds the concept of the genesis of the movement by including the forces that cause or underlie the motion that occurs. Forces are not visible; we can only infer forces by virtue of 'seeing' a motion take place or by other forms of measurement.

# 8

# LINEAR KINEMATICS APPLIED TO GYMNASTICS

*William A. Sands*

Taking a 'divide and conquer' approach and moving from simpler concepts to those with more complexity, we begin with linear kinematics. Linear kinematics refers to a description of motion that is in a straight line or nearly straight line (called curvilinear).

A simple example of an object moving in a straight line is a body falling straight down due to gravity. Linear or near-linear movement in gymnastics is seen most obviously during the vault run. We can also see linear movement in various phases of a skill.

An image of a vaulter during the takeoff phase from the vault board is shown in Figure 8.1. The 'stick figure' diagram of Figure 8.1 shows individual frames of movement as captured by high-speed video and then rendered by a computer and specialized software. Note that, as the apex of the head moves rightward and upward, the points are in a nearly linear arrangement and that the points get closer together as the movement proceeds.

Biomechanics of gymnastics uses this type of kinematic analysis. Most skills can be rendered in small enough segments of time that the small increments of motion are linear or nearly linear.

## 8.1 Distance and speed

As with other branches of science, biomechanics has special definitions of terms and the reader should be familiar with these terms and their applications.

When a body moves from one location to another, the *distance* of the movement is measured along the path that the body follows. Distance, by definition, is scalar: it does not involve direction and is a magnitude only.

If you express the distance covered relative to the time involved to cover that distance you have a new quantity called 'speed', in this case the 'average' speed. The term 'average' is added because we don't know if the speed changed at any point along the path of the movement; thus we have a simple average. Speed is described mathematically as in Equation 8.1. Average speed is also defined as the rate of change of position.

$$\textit{average speed}\left(\frac{m}{s}\right) = \frac{\textit{distance}(m)}{\textit{time}(s)} \tag{8.1}$$

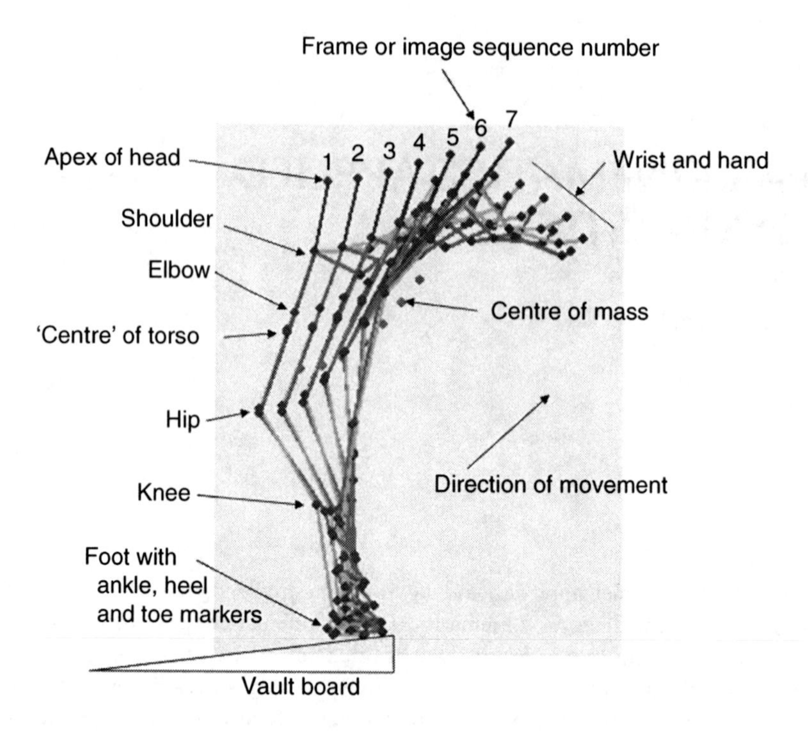


**FIGURE 8.1** Computer rendering of a gymnast taking off from a vault board towards the vault table. The motion is from left to right. Note that the paths of different joints and identified body parts are straight or slightly curved lines.

A simple example of speed can be seen when observing the vault run-up. If a vault has a run-up to the vault board of 15 metres (i.e., the run starts at 15 metres from the vault board) and the gymnast requires 2.5 seconds to run from the start to the vault board then the average speed would be:

$$average\ speed = \frac{distance}{time}$$

$$average\ speed = \frac{15m}{2.5s}$$

$$average\ speed = \frac{6m}{s}$$

Although the average speed can tell us about the run-up in its entirety, we know that the vaulter starts from a still position and then runs faster and faster until he/she reaches the vault board. The vaulter will often change his/her speed by speeding up in the beginning and then slowing down again just before board contact (Sands, 2000d; Sands & Cheetham, 1986; Sands & McNeal, 1995; Sands & McNeal, 1999).

Figure 8.2 shows the average speeds of a sprint test conducted on talent-identified female gymnasts of 9–12 years of age (Sands & McNeal, 1999). Note that the average speeds at the various distances changed. Although the overall average speed for this test was 5.5 m/s,

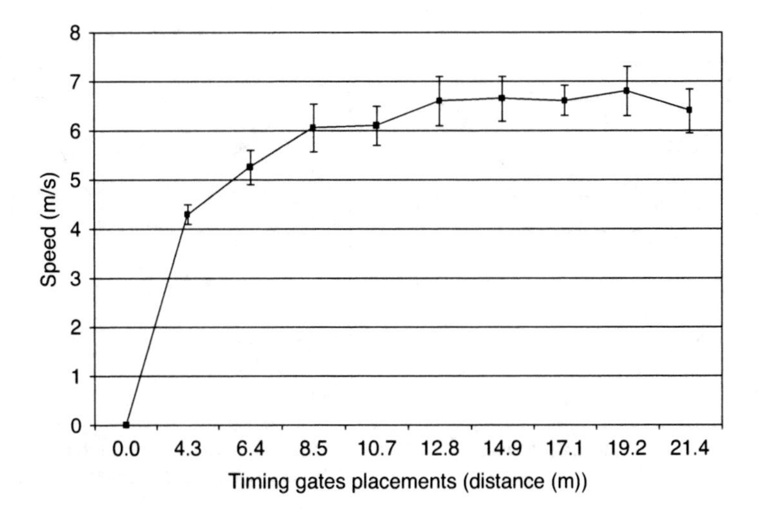


**FIGURE 8.2** Speed vs distance in gymnastics sprints.

we can see that there were probably very few, if any, recorded speeds at the average value. Moreover, the average value does not give a good idea of how the run-ups were actually performed by this group.

The information shown in Figure 8.2 leads us to consider how often we might want to sample the gymnast's movement in order to get a better idea of the subtler characteristics of his/her skill performance. Some movements that occur slowly may be analysed with relatively few samples of the movement while very rapid movements such as a tumbling or a vaulting takeoff may require analyses with samples acquired every thousandth of a second or faster.

## 8.2 Displacement and velocity

Most lay people find the difference between the concepts of speed and velocity to be trivial. However, in biomechanics and physics the difference between the two is important. Speed, like distance, is a scalar quantity, which means that the value has a magnitude only – direction is not considered. Velocity is a vector quantity and includes a direction as well as magnitude. You can think of the speed of a movement as the magnitude of a velocity while the direction must also be determined or considered in order to talk about velocity.

An example of the difference between the two concepts can be brought home by looking at a 400-metre run. In terms of the distance over which the runner goes if he/she starts and finishes at the starting line (same place), then the runner has covered a distance of 400 metres. If we know the time it took to cover the 400 metres then we can calculate an average speed. However, if the 400-metre runner starts and stops in the same place and we just look at the overall movement of the runner, then we would conclude that his/her *displacement* was *zero* metres. Displacement is a change of position taking into account direction as well as distance. The idea of the difference between distance and displacement, speed and velocity can be further illustrated by a floor exercise routine and the movement paths of the gymnast around the floor exercise area, Figure 8.3.

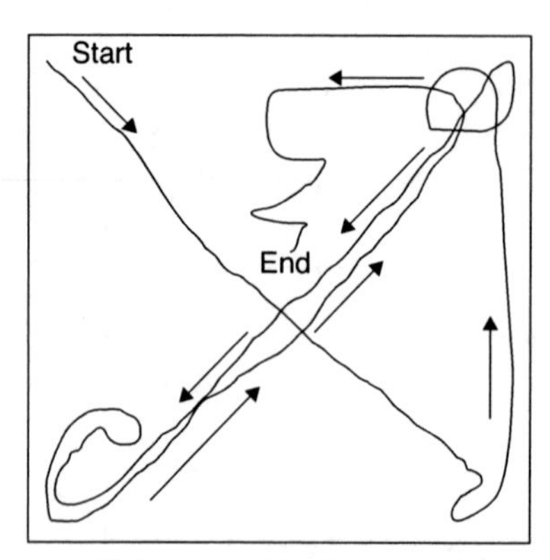


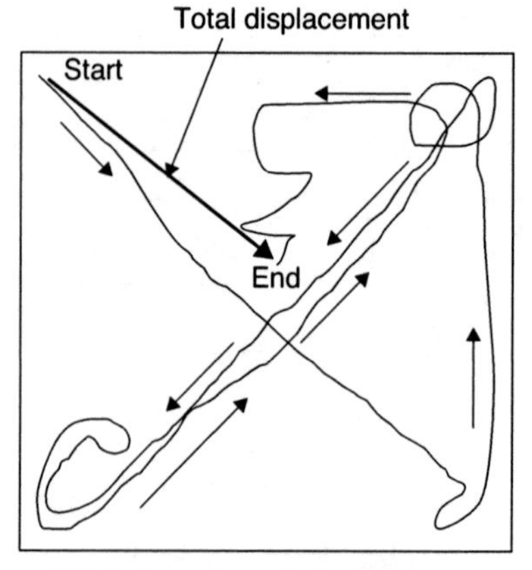


**FIGURE 8.3** Examples of distance and displacement. Each figure describes the path of a gymnast through a floor exercise routine. The left figure shows distance as measured along the entire path the gymnast follows, while the right figure shows the total displacement – the straight-line distance and direction from the start to the finish.

Average velocity is the quotient of the change in displacement and the change in time. See Equations 8.2 and 8.3.

$$\textit{average velocity}\left(\frac{m}{s}\right) = \frac{\textit{displacement}(m)}{\textit{time}(s)} \tag{8.2}$$

$$\textit{average velocity}\left(\frac{m}{s}\right) = \frac{\textit{final position}(m) - \textit{initial position}(m)}{\textit{final time}(s) - \textit{initial time}(s)} \tag{8.3}$$

It may seem that the idea of deciding what 'counts' as an initial position and a final position could be arbitrary. The biomechanist determines initial and final positions and initial and final times with considerable care in order to ensure that he/she is examining the portion of movement that is of interest. Moreover, we again confront the problem that an average speed or velocity probably provides limited insight in examining any movement. When looking closely at speed and velocity we tend to divide the movement into smaller and smaller time segments as a means of determining an 'instantaneous' speed or velocity. The term instantaneous refers to small segments of time and thus movement, and helps break a movement into smaller segments that reveal greater detail. An example of this process is shown in Figure 8.1.

## 8.3 Acceleration

A gymnast's ability to be explosive in his/her movements is often dependent on the ability to accelerate. Although rapid movement may be desirable, the gymnast is constrained by time limits of foot contact with the floor or apparatus, the correct moment for a release or other movements during a swing, and other constraints.

The rate of change of velocity is called acceleration. Average acceleration is shown mathematically in Equation 8.4.

$$average\ acceleration\left(\frac{m}{s^2}\right) = \frac{final\ velocity\left(\frac{m}{s}\right) - initial\ velocity\left(\frac{m}{s}\right)}{final\ time(s) - initial\ time(s)} \tag{8.4}$$

Clearly, Equation 8.4 shows that if the velocity is constant (so final velocity = initial velocity), then their difference equals zero, and thus there is no change in velocity and no acceleration.

Interestingly, an object at rest and an object in a straight-line uniform (i.e., unchanging) motion have equal accelerations – zero.

As in the case of speed and velocity, acceleration can also be determined as an 'instantaneous' value. If you look more closely at Equation 8.4, it is apparent that there can be positive, negative and zero accelerations, based on the values in the numerator.

When motions occur in a straight line, the positive and negative values provide little problem and people commonly refer to a positive acceleration as simple 'acceleration' and a negative acceleration as a 'deceleration'. In biomechanics, however, the word 'deceleration' is seldom used and the positive and negative designations are preferred. When the motion is not in a uniform straight line, as when a body moves in one direction and then reverses and moves in the opposite direction, we can face the problem of the interaction of direction and change in velocity. For example, a body may be moving in a positive direction (e.g., to the right) but slowing down, indicating a negative acceleration and vice versa. Keeping track of the positive and negative natures of direction and acceleration can prove difficult.

## 8.4 Linear kinematics units of measurement

Linear kinematics uses units that are relatively familiar to most people. Distance and displacement are measured in feet, inches, metres, centimetres, kilometres, miles and so forth. Time is measured in seconds, minutes, hours, days and so on. Logically, if a displacement is known, let's say in metres, and we divide that distance by the time required to cover it, we get velocity in metres per second. Other familiar speed and velocity measurements are feet per second, miles per hour, and kilometres per hour. The intuitive nature of these units often breaks down when we come to acceleration.

As you look at Equation 8.4, you should note that the number is the difference between final and initial velocities. The denominator is in seconds. Thus, average acceleration is measured in metres per second per second, or feet per second per second; this can also be expressed as metres per second squared, and so on. All of these are units of average acceleration.

When translated to simpler language, acceleration is a measure of the rate of change of velocity. For example, the speedometer in your car shows your current velocity. If you suddenly step on the gas and begin to move forward faster you'll see the speedometer needle or digital display show a rapid movement or a rapid change of digits. The speed of the movement of the speedometer is a measure of acceleration. If you suddenly hit the brakes while moving forward you'll note the speedometer shows a decreasing velocity. The more rapidly the speedometer changes, the larger the negative acceleration – you're slowing down faster.

All of these aspects of linear kinematics are shown in Figure 8.4. It shows a stroboscopic view of a gymnast performing a Roche vault (top image). The multiple images can provide a great deal of information about the gymnast's performance. The middle section of Figure 8.4 shows the gymnast's body rendered by a computer to simplify the motion and provide yet

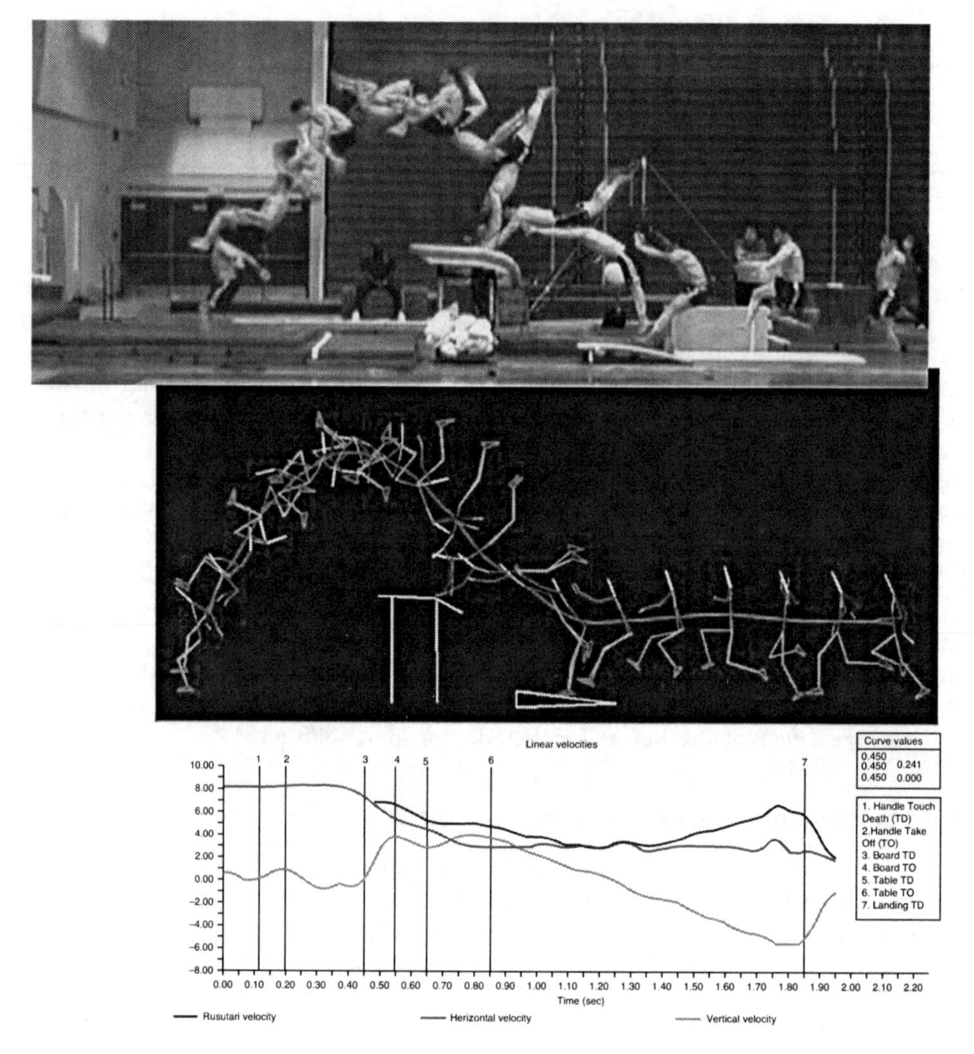


**FIGURE 8.4** Roche vault performed by an elite male gymnast. Note that the uppermost figure shows the performance as seen in a typical video format. The middle image shows a computer rendering (stick figures) along with the path of the centre of mass. Finally, the lower image shows the velocities of the centre of mass in the horizontal, vertical and resultant directions, phases of the skill and durations.

another view of the skill and shows the more linear component of movement that is the path of the centre of mass. Finally, the lower segment of Figure 8.4 shows the horizontal, vertical and resultant velocities of the centre of mass of the gymnast through the skill (Cormie *et al.*, 2004).

## 8.5 Frames of reference

The determination of the frame of reference within which the motion of the body and its segments occurs is an important part of the motion analysis process. Indeed, the frame of reference is necessary as the background on which motion is 'mapped'. Therefore, the frame of reference must be precisely determined in order to maintain precision in further analyses.

The frame of reference is defined by a point of reference and axes relative to which the motion is described (Figure 8.5). The gymnast's change in location can be described relative to a stationary environment $R_O$(O, xO, zO) such as the apparatus, for example, or relative to another arbitrarily defined reference $R_G$(G, xG, zG) or $R_S$(S, xS, zS), so that movements of specific points can be referenced to a fixed external 'framework' such that position, displacement, velocity and acceleration can be determined. For example, the movement of the athlete's centre of mass must occur within and about some frame of reference in order to have any physical meaning.

The three spatial axes – horizontal (xO), vertical (zO) and mediolateral (yO) – and the frontal, sagittal and transverse planes are commonly designated as the stationary axes and planes of reference for describing the direction of the body's motion in three-dimensional space. Whenever the orientation of the body's segments is moving in space, the spatial axes do not change and are fixed relative to the ground or relative to an observer. This stationary frame of reference is appropriate for analysing the trajectories of the body and its segments during gymnastic skills executed on the different kinds of apparatus. When performing aerial skills, the movement can be analysed in terms of accuracy of the segmental motion relative to the centre of gravity G during the flight phase. When the biomechanical purpose is to describe

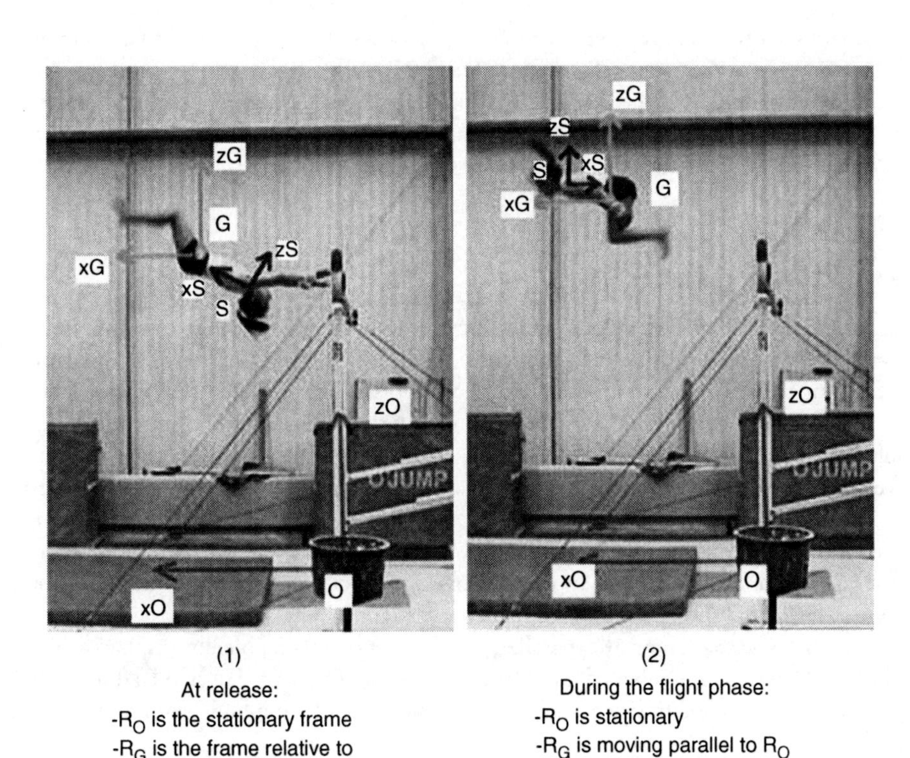


**FIGURE 8.5** Frames of reference usually used for analysing the motion of the body and its segments while performing gymnastic skills.

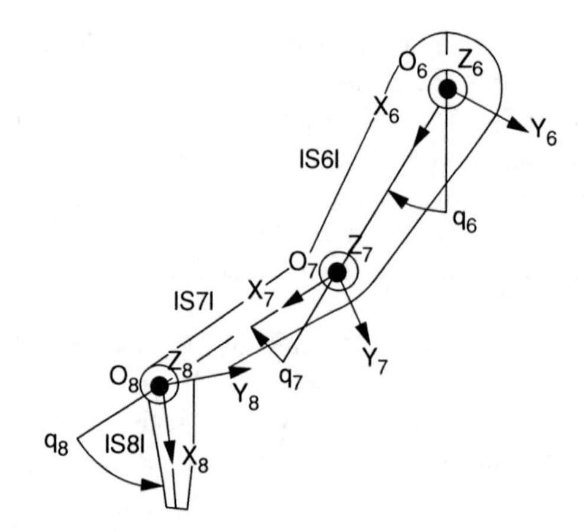


**FIGURE 8.6** Two-dimensional local frames of reference used for analysing segmental motions relative to the centre of joint. The lower limb motion is described with $q_8$, $q_7$ and $q_6$ relative respectively to each local frame: $R_8$ ($O_8$, $X_8$, $Y_8$); $R_7$ ($O_7$, $X_7$, $Y_7$); and $R_6$ ($O_6$, $X_6$, $Y_6$).

the segmental motion relative to the centre of gravity, it will be appropriate to use a frame of reference relative to the centre of gravity defined by $R_G$(G, xG, zG).

In other cases in which the description of the segmental motion relative to another segment is required, it is appropriate to define a moving frame of reference by the three principal axes (mediolateral, longitudinal and anteroposterior axes) passing through the centre of a joint (Figure 8.6).

## 8.6 Vectors and scalars

Thus far, we have included velocity and direction and mentioned the terms 'vector' and 'scalar', but we haven't fully defined the terms. A scalar has been discussed as the magnitude of something with no concern for direction. A vector is more complicated. A vector includes both magnitude and direction, and most importantly the magnitude and direction are simultaneously considered.

Velocity, acceleration, and many other values in biomechanics are vector quantities. Vectors are often visually depicted as an arrow from a point on the body at the beginning of the motion under consideration. The length of the arrow is scaled to the magnitude of the vector quantity and the direction in which movement occurs is depicted by the direction of the arrow. Figure 8.7 shows some examples of vectors.

Figures 8.7 and 8.8 show the parallelogram method for graphically displaying vector resolution. There are a number of interesting properties of vectors that will lead us to flight trajectories of bodies free in space, such as the flight phase of a tumbling skill, dismount and so forth.

A trajectory is shown in Figure 8.4 for a vault sequence. Note that the red line that follows the gymnast at about hip level depicts the path of the centre of mass during support and airborne phases. The path of this line is a trajectory. Vectors can be used to determine the nature of flight trajectories. Moreover, knowing the horizontal and vertical component velocity vectors can tell you most things you need to know about the subsequent flight trajectory.

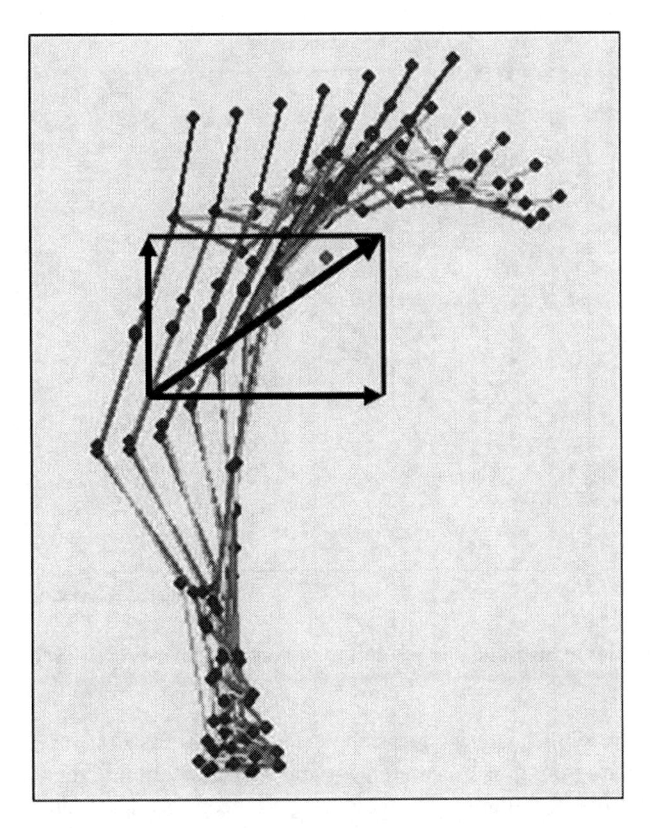

**FIGURE 8.7** Estimated horizontal, vertical and resultant velocity vectors of a vault takeoff. The horizontal velocity is greater than the vertical velocity due to the rapid run prior to the vault board contact. The vertical velocity component is shown as the vertical arrow. Both horizontal and vertical velocities have their origins in this case at the centre of mass of the gymnast. The angled vector is the resultant velocity of both the horizontal and vertical component velocities.

Figure 8.8 shows the vertical component velocity as a vector A, the horizontal component velocity as a vector *B* and a resultant (the actual resultant velocity of the object) as the vector *R* (resultant).

Traditionally, the angle of the resultant from the horizontal is called the angle of launch or projection and is denoted by the Greek letter $\theta$ (theta). The angle between lines A and B is a right angle and thus a great deal can be determined about the characteristics of the diagram of vectors in Figure 8.8 by knowing a little trigonometry. For example, if we know the horizontal and vertical component velocities we can determine the resultant simply by invoking the Pythagorean theorem in Equation 8.5.

$$R = \sqrt{(A^2 + B^2)} \tag{8.5}$$

Where: $R$ = resultant velocity
$A$ = vertical component velocity
$B$ = horizontal component velocity

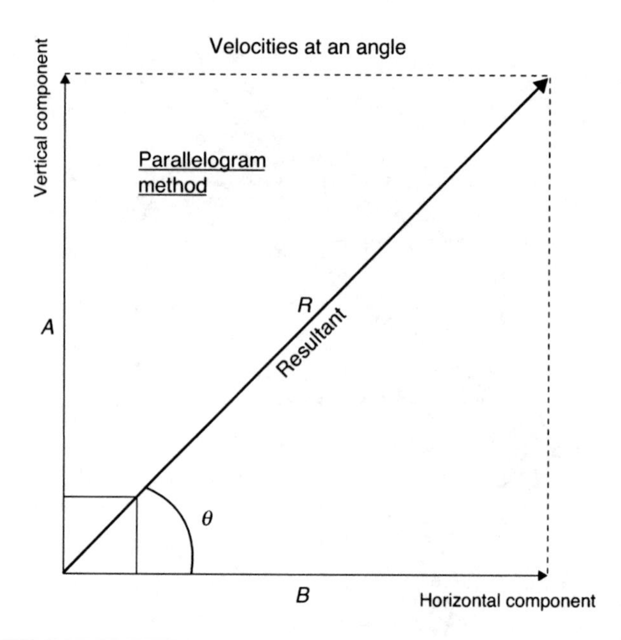


**FIGURE 8.8** Velocities at an angle and resolution of perpendicular velocity vectors.

More frequently, we estimate the location of the centre of mass (covered later), we know the resultant velocity from direct measurement and we can determine the angle of launch by observation of the first few frames of video during a takeoff or release. From this knowledge and some elementary trigonometry we can determine the horizontal and vertical component velocities (Equations 8.6 and 8.7).

$$\textit{vertical component velocity (line A)} = R\sin(\theta) \tag{8.6}$$

Where: R = resultant velocity
sin($\theta$) = the sine of the angle θ (theta)
A = the vertical component velocity shown in Figure 8.8

$$\textit{horizontal component velocity (line B)} = R\cos(\theta) \tag{8.7}$$

Where: R = resultant velocity
cos($\theta$) = cosine of the angle θ (theta)
B = horizontal component of velocity shown in Figure 8.8

To calculate the angle $\theta$, you need to know the horizontal and vertical component velocities and another trigonometric function called an arctangent. The arctangent of a number *x* is simply the angle whose tangent is *x*. Equation 8.8 shows the method for determining the angle of launch when the two component velocities are known.

$$\theta = \arctan\frac{A}{B} \tag{8.8}$$

Where: $\theta$ = angle between horizontal component of velocity and resultant
A = vertical component velocity
B = horizontal component velocity

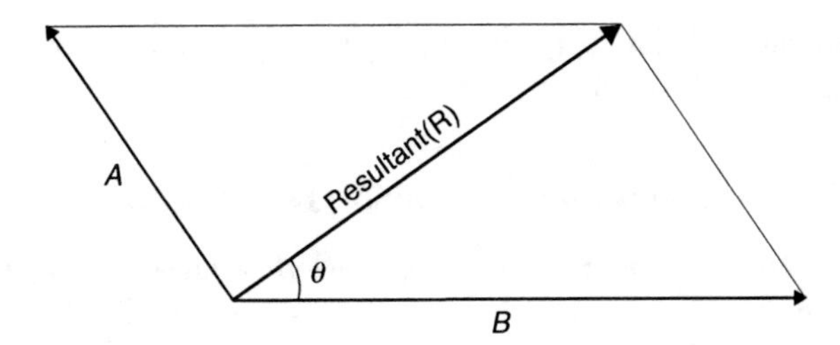


FIGURE 8.9 Velocities at a non-perpendicular angle and the resultant velocity.

Of course, not all resultant vectors can be reduced to two vectors at right angles and a resultant velocity in the middle (Figure 8.9). In the situation where the components are not at right angles the resolution of these component velocities and determination of the resultant velocity requires a different approach and a trigonometric identity called the cosine rule. Equation 8.9 shows the means of calculating a resultant where the angle $\beta$ (beta), the angle between vectors A and B, is not a right angle.

$$resultant(R) = \sqrt{A^2 + B^2 + 2AB\cos\beta} \tag{8.9}$$

The direction in which the resultant acts can be determined by Equation 8.10.

$$\theta = \arctan\frac{A\sin\beta}{B + A\cos\beta} \tag{8.10}$$

Both Figures 8.8 and 8.9 show depictions of a resultant velocity vector and its components. Equations have been listed such that components can be calculated from a known resultant and a known angle of projection. Or one can solve for the resultant by knowing the two components and the angle of projection. The figures also show completed parallelograms, which include lines that parallel the A and B vectors. These parallel lines that complete a parallelogram can be used to graphically solve for any of the components and the resultant. Figure 8.10 shows an example of the 'graphical method' that is used to solve for a resultant or the components. The horizontal component is 7 m/s, the vertical component is 3 m/s, the

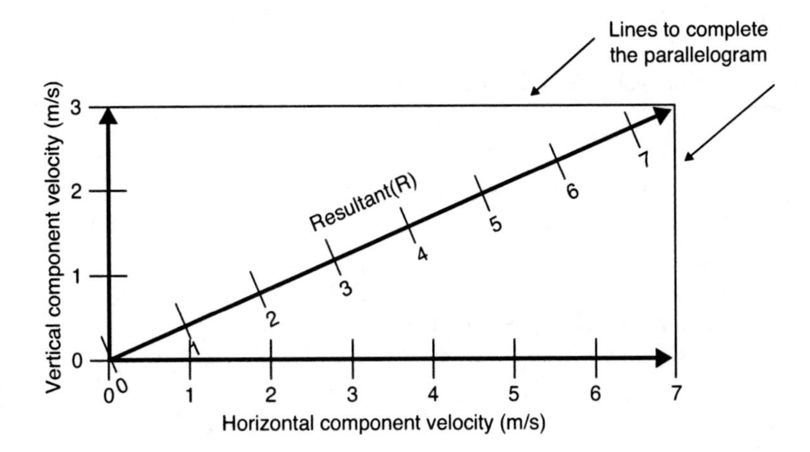


FIGURE 8.10 Graphical method of resolving vectors, and the resultant.

resultant is approximately 7 plus approximately 1/2 m/s based simply upon visual inspection. The calculated answer for the resultant is 7.6 m/s.

## 8.7 Taking flight: the kinematics of falling bodies and trajectories

Gymnasts often fall, and it is important to attempt to catch them and reduce injury (i.e., spotting). Interestingly, we can calculate the velocity at any point in a gymnast's falls, the time required to fall from some known height to the floor or any point between and given a certain launch angle and velocity (i.e., takeoff) we can determine nearly all aspects of the flight of the gymnast. These kinds of calculations become important in things such as spotting. For example, if a gymnast unexpectedly falls from a known height, can a human spotter intervene and catch the gymnast before he/she hits the ground or apparatus and perhaps becomes injured?

Sands (1996, 2000a, 2000c) described the problem of catching a falling gymnast when the physical principles of a falling body in the Earth's gravitation are coupled with the limited ability of humans to react to a stimulus through the three reaction stages of stimulus identification, response selection and response programming. These three stages occur within the brain and perceptual systems; then you have to consider movement time to some effective position to catch the gymnast. It turns out that in many situations in which a gymnast falls, catching him/her is nearly impossible, owing to the combination of acceleration of the fall due to gravity and the limited ability of humans to process and act on the information they see.

How can we be so sure of the information in the previous paragraph? The principles of freely falling bodies and trajectories have been known for centuries. Let's begin with a body falling straight down from some height. Because we live on the Earth with a nearly uniform gravitational field exerting a force (on everybody on its surface and in space around the planet) that is predictable and thus known, we can characterize a great deal about a gymnast's fall.

Three equations are all that are necessary to handle the characteristics of a freely falling body.

Velocity at any moment of the fall = initial velocity + acceleration × time

$$v = u + at \tag{8.11}$$

Distance of a fall = initial velocity × time + ½ acceleration × time$^2$

$$d = ut + at^2 \tag{8.12}$$

Velocity$^2$ = initial velocity$^2$ + 2 × acceleration × distance of fall

$$v^2 = u^2 + 2ad \tag{8.13}$$

Solving for time of the fall:

$$d = ut + \frac{1}{2}at^2 \tag{8.14}$$

Let's assume that initial velocity is zero; then:

$$d = \frac{a}{2}t^2 \tag{8.15}$$

$$t = \sqrt{(d/(a/2))} \tag{8.16}$$

In most gymnastics falls and trajectories the influence of air resistance is so small that it can be safely ignored. As an example of these equations and a falling gymnast, let's imagine that we have a gymnast standing on a balance beam (height = 1.25m). For simplicity let's say that the gymnast just steps off the beam and falls to the floor. We'll focus on the feet, although in most biomechanics settings the interest would be on the centre of mass (we'll come to that later).

$$\textit{time of the fall} = \sqrt{(d/(a/2))}$$

$$t = \sqrt{\frac{1.25}{\frac{9.806}{2}}}$$

$$t = \sqrt{\frac{1.25}{4.903}}$$

$$t = 0.50 \textit{ seconds}$$

The calculations show that the time required for the feet to descend to the floor after stepping off the balance beam is approximately 0.50 seconds. Typical simple reaction time is around 0.25 seconds, just to see a light flash and push a button (Henry & Rogers, 1960; Hodgkins, 1963; Stein, 1998). Unfortunately, trying to figure out something that will be effective in rescuing the falling gymnast takes a lot longer than 0.5 seconds (Woodson *et al.*, 1992).

For another example, let's assume that we have videotape of a dismount from the horizontal bar. We can plainly see when the athlete reaches the peak of his dismount flight trajectory. If we count frames from this point to the landing on the floor we can determine the duration of the descent. Each frame in standard video is taken in 1/30th of a second (0.03333s) U.S. (NTSC standard) video and 1/50th of a second using European (PAL) video. Let's assume the descent of the gymnast required 24 frames at 1/30th of a second each. This means that the athlete required 0.8 seconds to fall from the peak of his dismount flight trajectory.

To determine how far he fell, or the height of the dismount, we estimate by using this information (at the peak of the trajectory the vertical component velocity is zero):

$$d = ut + \frac{1}{2}at^2$$

$$d = (0 \times t) + \frac{1}{2}(9.806) \times (0.8)^2$$

$$d = 4.903 \times 0.64$$

$$d = 3.14 \textit{ metres}$$

Now, let's determine how fast the gymnast is going (only in a downward direction) at landing. If the time of the fall is 0.8 seconds, then to determine an approximation of the impact velocity all we need do is multiply the time that the acceleration acts on the falling body. At this point we should identify where the constant 9.806 comes from. The acceleration due to gravity is 9.806 m/s/s, also written as 9.806 $m/s^2$. This means that the velocity of a falling body increases at a rate of 9.806 metres per second for each second in which it falls. Thus, the longer the time of the fall the greater the final velocity (disregarding air resistance). However, note that the value for the acceleration is constant or uniform. This is a very important idea.

$$v = u + at$$

$$v = 0 + (9.806 \times 0.80)$$

$$v = 7.8\ \frac{m}{s}$$

Thus, the athlete's velocity at impact is 7.8 m/s.

There is an important caveat to the equations for uniformly accelerated motion – these equations only apply to situations in which the change in velocity is constant. Running, jumping (not the flight phase), swinging and so forth do change velocity, but the change or the acceleration is not constant. Thus, using these equations for anything but 'uniformly' accelerated motion' is a serious error. Fortunately, gravity provides us with a readily available source of uniformly accelerated motion via freely falling bodies.

Gymnasts rarely fall straight down; they usually rise from a takeoff or release and then fall to the ground or descend to regrasp a bar. Gymnasts can jump or rise straight up and then fall straight down. However, usually there is some horizontal travel during the flight as well. Interestingly, and simplifying understanding, it doesn't matter whether the gymnast moves horizontally or not, the principles of this type of flight are the same – we call this type of flight a 'trajectory'.

Figures 8.11 and 8.12 show a trajectory from a tumbling somersault. This particular somersault also included an attempt at four twists. The sequence runs from left to right and also includes the latter phase of the round-off and the complete back handspring (flic flac) along with the somersault. The trajectory of interest is the path of the centre of mass. The centre of mass is shown as a single dot in Figure 8.11 and as the series of dots in Figure 8.12.

Both Figures 20 and 21 show the two-dimensional velocities of horizontal and vertical components and the resultant velocity. Note the correspondence between the velocity changes that occur in the components and resultant along with the changes in motion and the flight trajectory.

The path of the centre of mass in Figure 12 shows a little 'noise' or bumpiness that is largely due to digitizing errors, which are artefacts of the hand-digitizing process that took place from the raw video footage of three cameras. However, the general shape of the curve of the flight path of the gymnast is parabolic and we will refer to these two figures in the following. You can also see a flight trajectory in Figure 11 where the gymnast leaves the vault table and launches into the post-flight somersault phase of the vault. Once the athlete leaves the ground and undertakes flight, he/she is a projectile – just like a baseball, football or bullet.

Two things are of primary interest in characterizing a trajectory: time of flight (height) and horizontal range (horizontal distance of travel). We can learn the time required to reach the top of the flight trajectory by knowing the resultant velocity and the angle of projection.

$$\text{time}_{\text{up}} = \text{resultant velocity at takeoff} \times \sin\theta$$

$$t_{up} = R\sin\theta \qquad (8.17)$$

In order to know the total time of flight we also have to know the time from the peak of the trajectory to the landing.

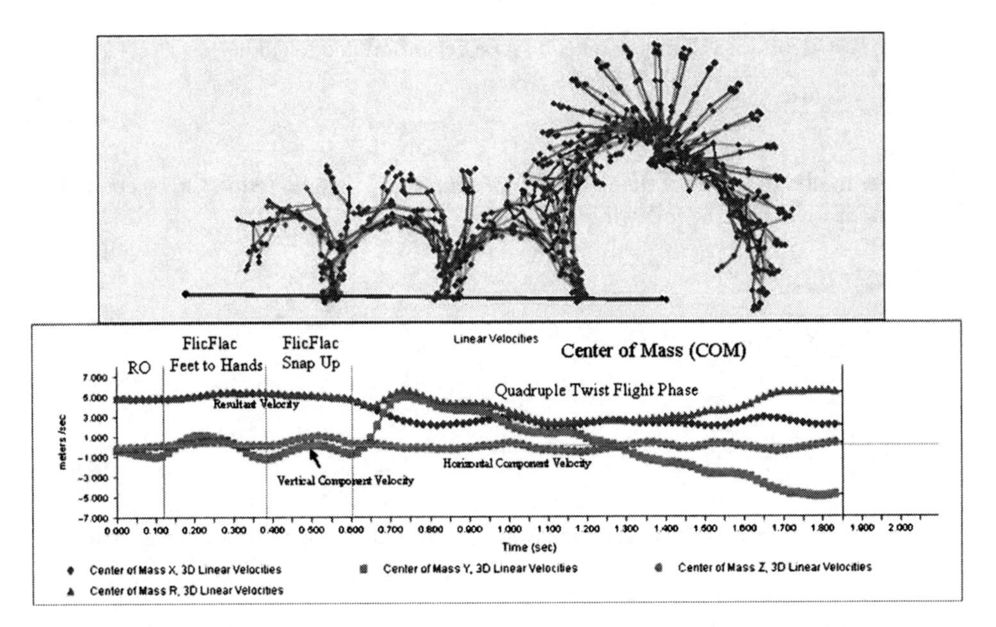


FIGURE 8.11 A stick-figure tumbling sequence of the snap-down of the round-off, back handspring (flic flac), and a high layout somersault with four twists. Note that the centre of mass in this figure is marked by a small triangle that is not connected to the stick-figure. The lower half of the figure shows the horizontal forward–backward and side-to-side, vertical and resultant velocities of the centre of mass.

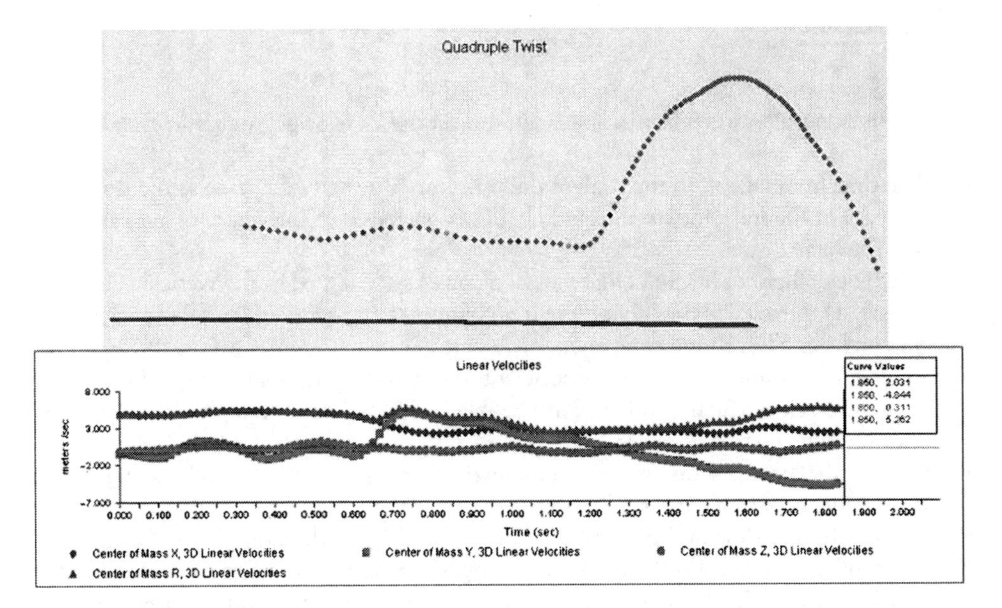


FIGURE 8.12 Centre of mass path. The stick-figure is removed from this image, leaving the movement of the centre of mass as shown by the small triangles. The rightmost movement of the centre of mass shows the large bump in the trajectory of the gymnast during flight.

time$_{down}$ = resultant velocity at takeoff / g (acceleration due to gravity)

$$t_{down} = \frac{R\sin\theta}{g} \tag{8.18}$$

In order to obtain the total time of flight we simply add the two terms. Or, by combining the equations we obtain the following:

$$T = t_{up} + t_{down}$$

$$T = \frac{R\sin\theta}{g} + \frac{v_{up}}{g}$$

$$T = \frac{R\sin\theta + v_{up}}{g} \tag{8.19}$$

An important caveat of the foregoing is that if the launch and landing are at the same level then:

$$t_{up} = t_{down}$$

Which can also be written as:

$$\frac{R\sin\theta}{g} = \frac{v_{up}}{g}$$

And, total time can be determined by the following:

$$T = \frac{2R\sin\theta}{g} \tag{8.20}$$

These relationships, when launch and landing are at the same height, indicate several things:

- The time from takeoff to the peak of the trajectory is exactly the same as the time from the peak of the trajectory to the landing. (This assumes that the centre of mass rises and falls the same distance.)
- The time of flight or height of flight are completely determined by the vertical velocity of takeoff. Thus, you will note that there is no term in the above equations for time of flight that includes a horizontal component of velocity.
- In order to maximize time of flight or height, the gymnast must maximize his/her vertical takeoff velocity, which involves considerable technique and strength/power considerations leading up to the takeoff.
- Once in the air, the gymnast can do absolutely nothing to alter the parabolic flight path that his/her centre of mass is following.
- The height and time of flight and the horizontal range of the projectile are completely independent of each other in terms of the physics. However, we know that it is considerably harder to go high than far in a jump. The negative acceleration that is caused by gravity can be thought of as ‘robbing’ the vertical component velocity of 9.806 m/s in each second of flight. Thus, if you leave the ground with a vertical component velocity of 9.806 m/s, you will reach your trajectory peak in 1 second.

Horizontal displacement becomes more problematic to calculate because we must determine the time of flight (determined by vertical velocity at takeoff) and the horizontal component velocity. Both of these values can be decomposed from the resultant; however, until now we've not shown the horizontal component velocity as calculated from a known resultant velocity and the angle of projection (θ).

Starting with Equation 8.24, we can substitute Equation 8.20 for time of flight. Then we must determine the average horizontal component velocity which is given by:

$$\textit{Horizontal component velocity} = R\cos\theta \tag{8.25}$$

Thus:

$$s = R\cos\theta \times \frac{2R\sin\theta}{g}$$

The equation above can be simplified by using a trigonometric identity

$$\sin 2\theta = 2\sin\theta\cos\theta$$

which then gives us:

$$s = \frac{R^2 \sin 2\theta}{g} \tag{8.26}$$

Equation 8.26 shows that the horizontal range ($s$) of a body or object is determined by the resultant velocity at launch and the angle of the projection. In Equation 8.26, $g$ is a constant acceleration due to gravity and therefore not within the control of the gymnast.

Of course, Equation 8.26 presumes that launch and landing are at the same height. If this is not the case, the equation becomes more complicated, but the terms should be easily recognizable. The equation is simply a fancier way of using average horizontal component velocity and knowing the total time in the air.

$$s = R\cos\theta \times \frac{R\sin 2\theta + \sqrt{(R\sin\theta)^2 + 2gh}}{g}$$

The equation above can be reduced to:

$$s = \frac{R^2 \sin\theta\cos\theta + R\cos\theta\sqrt{(R\sin\theta)^2 + 2gh}}{g} \tag{8.27}$$

In both circumstances – the same launch and landing positions, and different launch and landing positions – velocity of launch or takeoff is important. In the case of launch and landing at the same level, the optimum angle of launch or takeoff is 45 degrees measured from the centre of mass. In the case of launch and landing at different heights, the problem becomes one of optimization. When trying to achieve a maximum range or distance of flight, the following rules apply:

- Equal changes in either height of launch or speed of launch do not result in equal changes in the optimum angle or the horizontal range.
- The best angle for launch is always less than 45 degrees.

The preceding has involved an assumption that the takeoff and the landing occu level. Often this assumption can provide fairly close approximations of flight time However, there are circumstances where the gymnast launches from a position th bit higher than the landing (e.g., vaulting post-flight, dismounts from the appa Moreover, there are also circumstances when the takeoff point is below the landin such as a mount to the balance beam, uneven bars, parallel bars and so forth. In these determining the total time of flight is more complicated and the equation a bit more c

$$T = \frac{R\sin\theta + \sqrt{(R\sin\theta)^2 + 2gh}}{g}$$

In this case, $R$ again refers to the resultant velocity, $\theta$ the angle of launch, and acceleration due to gravity. The only additional term is $h$ which is the height of the from the landing point at the moment of launch. One should keep in mind that the referred to here is often the height of the centre of mass, not simply the height of the from which the launch occurred. Various body position distortions can occur and nak appraisals of takeoff and landing positions are often deceptive.

Thus far the primary characteristic of interest has been time. What about displacer Displacement occurs in two dimensions: forward–backward and up–down. However, we ally just concern ourselves with the horizontal range from launch to landing and the ver height of the trajectory. Height of flight is the easiest characteristic to determine: all vert displacements are assumed to begin at the point of launch and the calculation is showr Equation 8.22.

$$\textit{height of flight} = \frac{(R\sin\theta)^2}{2g} \qquad (8.2$$

You can also think of the height of flight as:

$$H = \frac{v_{up}^2}{2g} \qquad (8.23)$$

Again, it should be obvious that the only things that determine the height of flight of any projectile are the negative acceleration due to gravity and the vertical velocity at launch. Although the athlete can perform a variety of airborne movements, (e.g., swinging arms, tucking the legs, arching, piking, etc.), the only thing under the athlete's control in achieving a high flight trajectory is his/her vertical velocity at takeoff or launch.

Horizontal displacement is simultaneously simple and complex. Because the horizontal component velocity and the vertical component velocity are completely independent, one need only know the horizontal component velocity of the object or body and the time of flight. As such, the object or body gets to travel in a horizontal direction only so long as it's in the air; gravity has no influence on horizontal velocity. Therefore:

*horizontal range = average velocity × time of flight*

Or:

$$s = vt \qquad (8.24)$$

Where: $s$ = horizontal range
$v$ = horizontal component velocity
$t$ = total time of flight

- For any particular height of launch, the greater the resultant velocity the more closely the best angle of launch approaches 45 degrees.
- For any particular resultant velocity of launch, the greater the height of the launch the lower the optimum angle (Hay, 1973).

For launch and landings that move from a lower to a higher position, such as a vault board takeoff to the vault table, the optimum angle of departure varies based on the intent of the landing on the higher surface. For example, do you want to land and stop (balance beam mount) or do you want to preserve as much horizontal component velocity as possible (vaulting)?

## 8.8 Conclusion

Linear kinematics provides a foundation for understanding much of the following material. You will see positions, displacements, velocities and accelerations throughout the remainder of this chapter. Moreover, in the next section on angular (rotational) kinematics, you will see how the linear ideas are only slightly modified and an additional thought process is to hold two thoughts in mind simultaneously, linear ideas and additional information about their rotation.

# 9

# ANGULAR KINEMATICS APPLIED TO GYMNASTICS

*William A. Sands and Patrice Holvoet*

When a gymnast moves from one position to another, he/she may do so by translation (linear movement), rotation about an axis and through an angle (angular movement) or both. Angular movement is 'science-speak' for 'rotation'. Somersaults, handsprings, giant swings, hip circles, pommel horse circles and so forth are all examples of skills that involve angular motion. Fortunately, there are angular analogues to linear motion, making the understanding of angular motion simply an extension of the concepts from linear motion.

Different sports describe angular movement using terminology that fits sporting contexts and usually has origins in long traditions. Diving and gymnastics refer to somersaults as forward (face leading) or backward (back of the head leading). However, in gymnastics a backward giant swing has the face leading and a forward giant swing has the back leading. The terminology in these cases appears contradictory. Diving uses both forward and backward somersaults along with 'inward' and 'gainer' to indicate which way the diver is facing relative to the diving board or platform during the takeoff.

If we add twisting to rotation terminology then the direction is referred to as right or left. A right twist has the right shoulder leading rearward or the head turns to the right. A left twist shows the opposite. The magnitude of twists and somersaults is usually simply described as single, double, triple and fractions thereof. Biomechanics uses a different approach to determining the direction and magnitude of rotation.

## 9.1 Angular motion

Although it is perfectly acceptable to talk about angular motion in terms of degrees, the more common approach used in biomechanics is to refer to angular motion in terms of radians. The reason for this is that there is a serious mathematical problem with degrees when the rotation involves more than one complete rotation. For example, if a diver completes two and one half somersaults, then he/she has performed 900 degrees of rotation. If we watch the diver rotate and standardize vertical as 0 degrees, then as the diver completes a single rotation his/her position changes from 359 degrees to 0 degrees very suddenly and creates a difficult problem for mathematics because of the continuous change in position with a discontinuous change in the mapping of that position onto a 360-degree circle.

While degrees are terrific for navigational purposes, they are much more problematic in a mathematical, geometric and trigonometric world. As such, there is an easy way to determine the rotation of something by taking advantage of properties of a circle. For example, we know that the circumference of a circle is given by the equation:

$$circumference = 2\pi r \tag{9.1}$$

Where: $\pi = \text{pi} \approx 3.1416$
– the ratio of the circumference of a circle to its diameter, $2r$.

We can use this to express an angle in terms of radians. One radian is defined as the angle that marks off a distance on the circumference of a circle equal to the radius of the circle. The whole circle therefore contains an angle of $2\pi$ radians. Since this is 360 degrees, one radian $\approx$ 57.3 degrees.

In the world of gymnastics we might say that a gymnast performed a one-and-one-half twisting double somersault. In degrees this would be a 540-degree-twisting 720-degree somersault, or a $3\pi$-twisting $4\pi$ somersault.

## 9.2 Angular speed and angular velocity

Angular speed and angular velocity are terms used similar to their linear counterparts. The angle through which a body moves in a period of time is called angular speed, if you're not concerned with direction, and angular velocity if you are.

$$angular\ speed = \frac{angular\ distance}{times} \tag{9.2}$$

If you're concerned with both the magnitude of the angular distance and the direction then you use angular velocity as the correct term, usually given by the Greek letter $\omega$ (lower-case omega).

$$\omega = \frac{\theta}{t} \tag{9.3}$$

For example: if a gymnast rotates counterclockwise 60 degrees in a giant swing (and you've defined clockwise as positive) in 0.45 seconds, then the result is:

$$\omega = \frac{-60\ degrees}{0.45\ seconds}$$

$$\omega = -133\ \frac{degrees}{seconds}$$

## 9.3 Angular acceleration

Angular acceleration works in the same way as linear acceleration. Angular acceleration is the rate at which angular velocity changes with respect to time. Or, angular acceleration is the final angular velocity minus the initial angular velocity per unit (or divided by) time. Angular displacements, velocities and accelerations can be both average and instantaneous values just like their linear counterparts.

$$\alpha = \frac{\omega_f - \omega_i}{t_f - t_i} \tag{9.4}$$

A gymnast who is travelling in a clockwise direction at 100 degrees/second at one point in a giant swing and then 0.6 seconds later is travelling at 140 degrees/second then the average acceleration of the gymnast would be:

$$A = \frac{\frac{140\ deg}{sec} - \frac{100\ deg}{sec}}{0.6\ sec}$$

$$A = \frac{\frac{66.7\ deg}{s}}{s} \quad or \quad \frac{66.7\ deg}{s^2}$$

As shown in Figures 8.7 and 8.8 it is customary to graphically depict motion as vectors representing magnitude and direction.

Angular motion is little different except it is difficult to graphically depict rotational motion on two-dimensional paper. Therefore, a convention for showing angular motion vectors has been derived, called the 'right-hand thumb rule'. In order to describe an angular motion vector you use your right hand with the fingers flexed in a fist and the thumb out in a sort of 'thumbs up' gesture. The direction of the fingers indicates the direction of rotation of the object while the thumb represents the direction in which the vector is drawn.

Since angular velocity, acceleration and other quantities are vector quantities, the right-hand thumb rule can be applied to all of these. Figure 9.1 shows the takeoff for a quadruple twisting backward somersault. Considering only the twist the gymnast is turning to the left, or the gymnast's face is turning toward the reader and away from the page. If we apply the 'right-hand thumb rule' to set up our description of the twist velocity it would look like Figure 9.2.

Angular kinematics has one more aspect that requires explanation. Although we've been concerned with angular motions so far, described and calculated as an angular change in

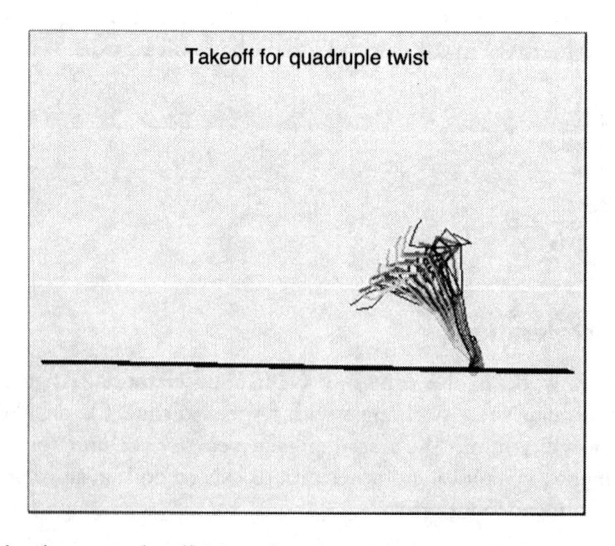


**FIGURE 9.1** Quadruple-twist takeoff. Note that the gymnast is turning so as to face to the left prior to the twist and then to face the reader during the pictured initial phase of the twist.

**FIGURE 9.2** The right-hand thumb rule. The direction of the fingers in the fist shows the direction of the turn, while the thumb shows the direction in which you would draw the velocity vector of the twist.

the position, velocity or acceleration of a body, there is another characteristic of angular motion, which is the linear speed or velocity of the object at any given point in the angular movement. For example, in a golf swing the head of the club, while in angular motion, also has a linear motion component that might apply to the head of the club as it strikes the ball on the tee.

Average angular speed is measured as the angular distance through which the body rotates in a period of time. For example, if an object is swinging about an axis through an arc AB then

$$v_{Tangential} = \frac{distance}{time} = \frac{arc\ AB}{t} \quad (9.5)$$

where $v_{Tangential}$ is the linear speed of any point on the object.

Since the angular distance is measured in radians, the number of radians is determined by dividing the arc *AB* by the radius, in this case *r*.

$$\omega = \frac{v_T}{r}$$

Rearranging:

$$v_T = \omega r \quad (9.6)$$

Finally, you can consider angular motion to be composed of two parts, both alluded to in the previous sections. One is a radial component which is the component that causes the object to rotate about an axis. If you swing a ball on a string in a circle above your head, the string is representative of the radial component. The string constrains the movement of the ball and, by virtue of making the ball constantly change direction, accelerates the ball.

If you release the string at any given moment, then the ball will initially fly away at a tangent (i.e., at right angles to the radius of the circling motion, i.e., the string) in the plane of the circling ball and string. The tangential component refers to the tangent of a circle, which is a line drawn at right angles to the radius and which touches the circumference of the circle at only one point. The tangential component of angular motion is important because it perfectly predicts in what direction an object will move once the radial component is eliminated. The radial acceleration is given by

$$\alpha = \frac{v_T^2}{r} \quad (9.7)$$

The term $v_T^2$ is the speed of the ball tangential to its circular path as it is swung around your head by the string. When the ball is released, the speed of the ball will be this term.

Moreover, we can determine that the flight path of the ball will be at a tangent to the circle in which it was spinning. The exact same principle applies to the direction of flight following a release on the horizontal bar or uneven bars from a giant swing. The rate at which the speed of the ball changes as it moves along its curved path is called the tangential acceleration and is calculated by:

$$\alpha = \frac{v_{Tf} - v_{Ti}}{t} \tag{9.8}$$

This information is particularly important in some sports or events such as the hammer or discus throw in track and field.

In gymnastics these ideas can be seen in the swing and release of a flyaway. Figure 9.3 shows a dangerous flyaway from the uneven bars, where the release occurred when the tangential motion of the centre of mass of the athlete caused her to move toward the bar and nearly strike it. Figure 9.4 shows a superior flyaway where the release and tangential flight path caused the gymnast to move away from the bar (Sands *et al.*, 2004b).

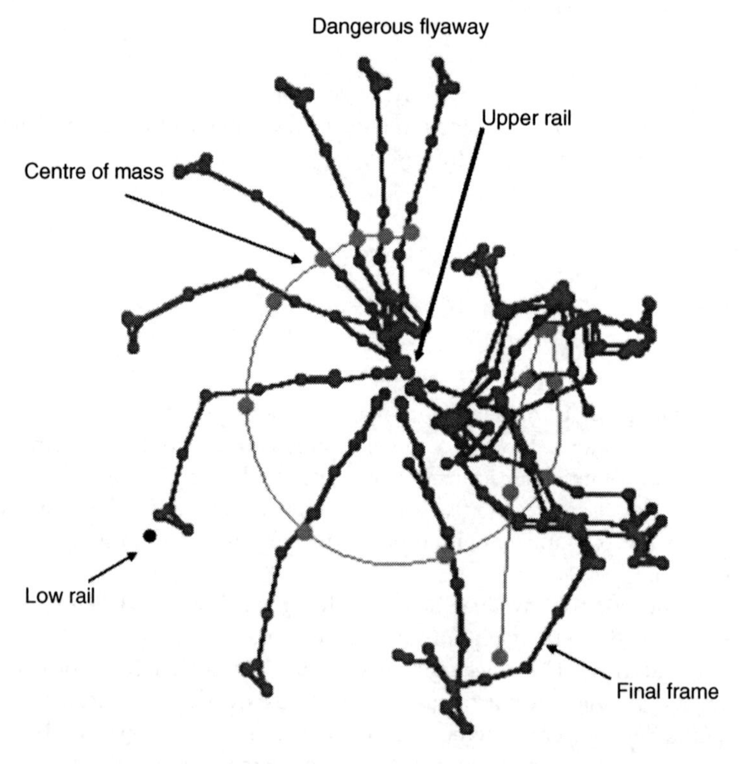


**FIGURE 9.3** Dangerous flyaway. The dotted line depicting the path of the centre of mass of the gymnast shows that she released too late and her trajectory took her towards the high bar of the uneven bars.

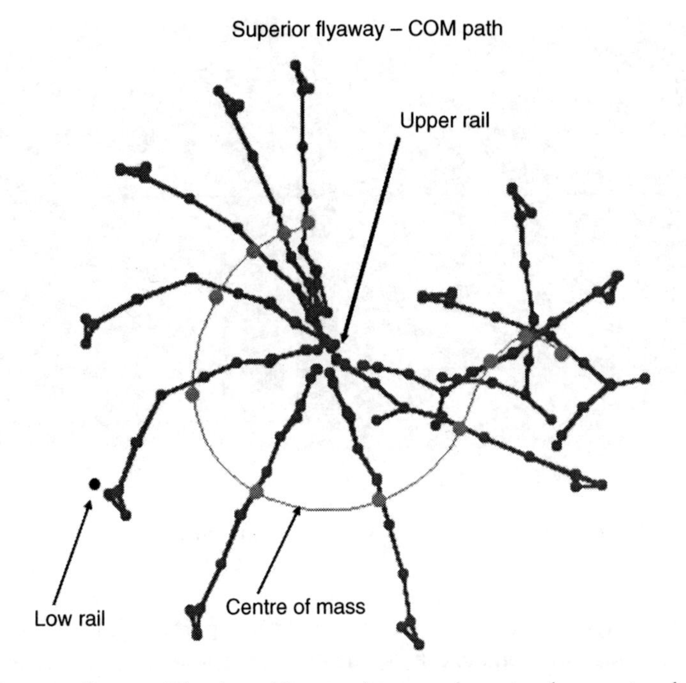


**FIGURE 9.4** Superior flyaway. The dotted line in this case shows a release point that resulted in the gymnast moving up and away from the high bar of the uneven bars in a flight trajectory that was both more effective and safer.

## 9.4 Application: understanding relations between angular and linear motions

Take, for example, the analysis of the forward swing on the high bar performed by a young gymnast. What about angular and linear velocities? Figure 9.5 shows that at the beginning of the downward phase ($t_0$=0s) the hip is located at a radius of gyration of 1.35m from the bar ($rHipt_0$) with values of angular and linear velocities equal to zero. When the gymnast is swinging just under the bar ($t_1$=0.56s), the hip has rotated through an angle ($\theta_1$) of 140° (or 2.44 rad) with a radius of gyration equal to 1.65m. When the gymnast reaches the front horizontal level of the bar ($t_2$=1s), the hip has travelled through an angle of 248.5° (or 4.33 rad) with a radius of gyration of 1.45m.

The axis of the angular velocity vectors $\omega_1$ and $\omega_2$ at $t_1$ and $t_2$ respectively is perpendicular to the plane of rotation and is along the bar. The direction of the angular velocity vectors along the bar is given by the right-hand thumb rule. The angular speeds $\omega_1$ and $\omega_2$ at $t_1$ and $t_2$ respectively are given by:

$$\omega_1 = \frac{\theta_1 - \theta_0}{t_1 - t_0} = \frac{140}{0.56} = 250°/s \quad or \quad 4.36\ rad/s$$

$$\omega_2 = \frac{\theta_2 - \theta_1}{t_2 - t_1} = \frac{248.5 - 140}{1 - 0.56} = 246°/s \quad or \quad 4.30\ rad/s$$

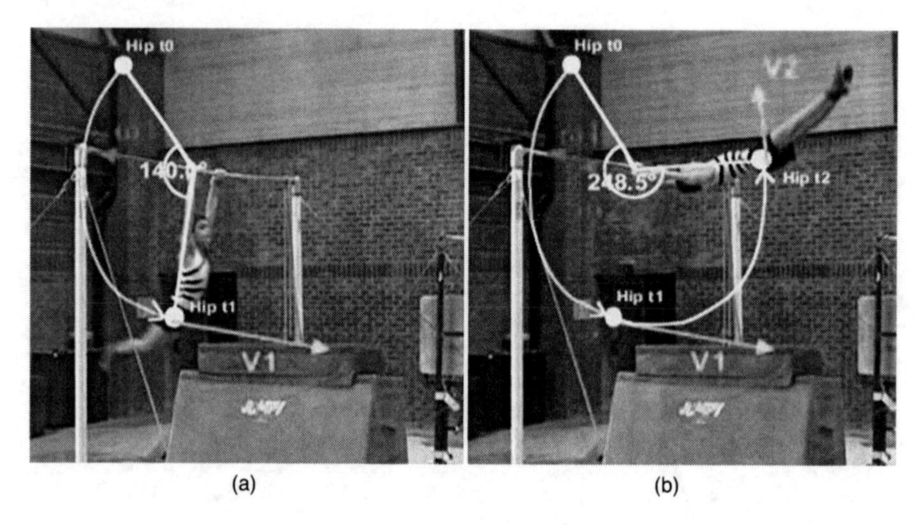


**FIGURE 9.5** Angular velocity and linear velocity of the hip during a forward swing around the high bar.

Knowing $\omega_1$ and $\omega_2$ and the radius of gyration of the hip from the bar at $t_1$ and $t_2$ respectively, the linear velocities $v_1$ and $v_2$ can be calculated as follows:

$$v_1 = rHipt_1.\ \omega_1 = 1.65 \times 4.36 = 7.19\ m/s$$

$$v_2 = rHipt_2.\ \omega_2 = 1.45 \times 4.30 = 6.23\ m/s$$

The direction of the linear velocities, vectors $v_1$ and $v_2$, is along the tangent line touching the curved path at $t_1$ and $t_2$.

The hip swinging action performed by the gymnast between $t_1$ and $t_2$ helps to avoid losing too much angular and linear speed during the upward phase.

Now let's determine how angular and linear accelerations are given during the swing.

Figure 9.6 shows that at the beginning of the upward phase ($t_1$=$t_{initial}$), the angular speed of the gymnast is equal to $\omega_1$. At $t_2$ ($t_2$=$t_{final}$) the gymnast reaches the front horizontal level of the bar with an angular speed equal to $\omega_2$. The average angular acceleration ($\alpha$) acting during the upward phase is calculated as follows:

$$\alpha = \frac{\Delta\omega}{\Delta t} = \frac{\omega_2 - \omega_1}{t_2 - t_1} = \frac{4.30 - 4.36}{1 - 0.56} = -0.13\ rad/s^2 \quad or \quad -7.95°/s^2$$

That negative angular acceleration is responsible for the deceleration of the angular motion of the gymnast during the upward phase.

It appears similarly in Figure 9.6 that at the beginning of the upward phase ($t_1$=$t_{initial}$), the linear speed of the gymnast is equal to $v_1$. At $t_2$ ($t_2$=$t_{final}$) the gymnast reaches the forward horizontal level of the bar with a linear speed equal to $v_2$. The average linear acceleration ($a$) acting during the upward phase is constructed with the vector $V_2$–$V_1$ (Figure 9.6a) and is calculated as follows:

$$\alpha = \frac{\Delta V}{\Delta t} = \frac{V_2 - V_1}{t_2 - t_1} \tag{9.9}$$

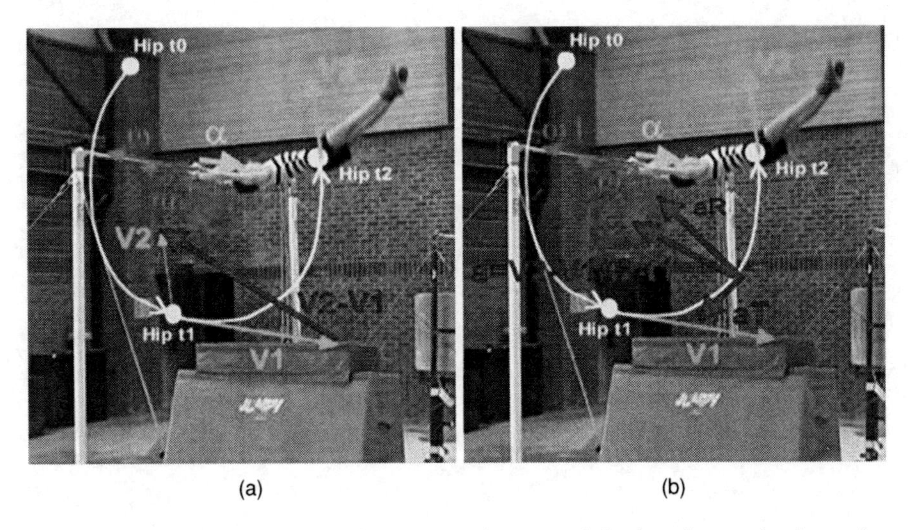


FIGURE 9.6 Angular acceleration and linear acceleration of the hip during the forward swing around the high bar.

The vector drawing of the linear acceleration can be considered as the vector resulting from the addition of a tangential component vector ($aT$) and a radial component vector ($aR$) (Figure 9.6b).

The vector $aT$ determines the tangential acceleration and is acting along the tangential line passing through each point of the curved path. The vector $aR$ determines the radial or centripetal acceleration and is acting along the radial line connecting each point of the curve path to the centre of rotation.

Remember that the linear acceleration ($a$) of a point rotating with a given angular acceleration ($\alpha$) depends also on its radius of rotation ($r$). The greater the radius of gyration, the greater the linear acceleration of the rotating point. The relationship between linear and angular acceleration of a rotating point is given as follows:

$$a = r\alpha \tag{9.10}$$

*With:* $a$ *in* $m/s^2$
$r$ *in* $m$
$\alpha$ *in* $rd/s^2$

If we suppose that during the downward phase of the swing (Figure 9.6) the radius of rotation of the hip is constant and is equal to $rHipt_1$, the average linear acceleration of the gymnast between $t_1$ and $t_0$ is given by

$$a = r\alpha \quad with \quad r = 1.65\ m$$

$$\alpha = \frac{\omega_1 - \omega_0}{t_1 - t_0} = \frac{4.36}{0.56} = 7.78\ rd/s^2$$

$$a = 1.65 \times 7.78 = 12.83\ m/s^2$$

During the downward phase of the swing the linear acceleration of the gymnast is equal to 1.3 times the acceleration due to gravity.

## 9.5 Conclusion

Angular kinematics follows many of the same conventions as linear kinematics.

However, angular kinematics often involves several aspects of physical laws occurring simultaneously. A swing, for example, has both a tangential and a radial component. As we move into linear and angular kinetics, the importance of these differences will become even more pronounced.

# 10

# LINEAR KINETICS APPLIED TO GYMNASTICS

*William A. Sands*

Kinematics is a description of motion: position, displacement, velocity and acceleration. There is no reference in kinematics to the underlying causes of motion.

Kinetics is the area of biomechanics where the source of motion is studied. The source of motion is a force.

Unlike the visibility of position, displacement, velocity and acceleration, forces are invisible. We can only infer that a force is present when we see or measure a body's acceleration. There are a few terms that should be defined before proceeding further. And one of them is important to the idea that an acceleration or a tendency to accelerate must be present in order for there to be a force.

## 10.1 Inertia

Inertia is the property of a body of remaining in one place or continuing in uniform straight-line motion unless acted upon by an outside force. It may seem counterintuitive, but an object at rest (i.e., not moving) and an object in straight-line uniform motion are actually examples of the same thing – inertia.

Often, the word 'reluctance' is used when describing inertia. It's important to realize that the biomechanist is not ascribing a property of consciousness to an object and that its reluctance does not come from not 'wanting' to move, but there are few words in English that suffice to describe this phenomenon without collapsing to circularity. Inertia is proportional to mass, so a more massive object, whether at rest or in straight-line uniform motion, exhibits greater reluctance to move than a less massive object.

## 10.2 Mass

Mass is the quantity of matter in an object. Weight and mass are constantly confused. Weight is a measure of force, usually the force of gravity pulling on your mass. You can measure your weight using a scale. However, when you are in a freefall situation, such as while orbiting the Earth, you are weightless (or nearly so in microgravity), but you still have the same mass as when you were on the surface of the Earth. The distinction of weight and mass is one of the reasons we prefer to use 'centre of mass' rather than 'centre of gravity' when describing properties of a body or object; however, here on Earth, the ideas are interchangeable. The Imperial/

English measurement system unit for mass is the 'slug'. It comes from the term 'sluggish' and implies the 'reluctance' or resistance to motion that is described by the term 'inertia'. In the metric system, the fundamental unit of mass is the kilogram.

## 10.3 Force

Forces change a body's state of motion, either moving it from rest (accelerating it), changing its direction (such as radial acceleration discussed in angular kinematics) or changing the velocity of an already moving object (i.e., also an acceleration).

A force is defined as a push, pull or tendency to distort. The push and pull can be seen in the acceleration that results. The 'tendency to distort' idea appears in situations where there is a force applied to something but the force is insufficient to actually accelerate the object. This might occur when the gymnast is trying to press to handstand (Figure 10.1) but is unable to do so; or the

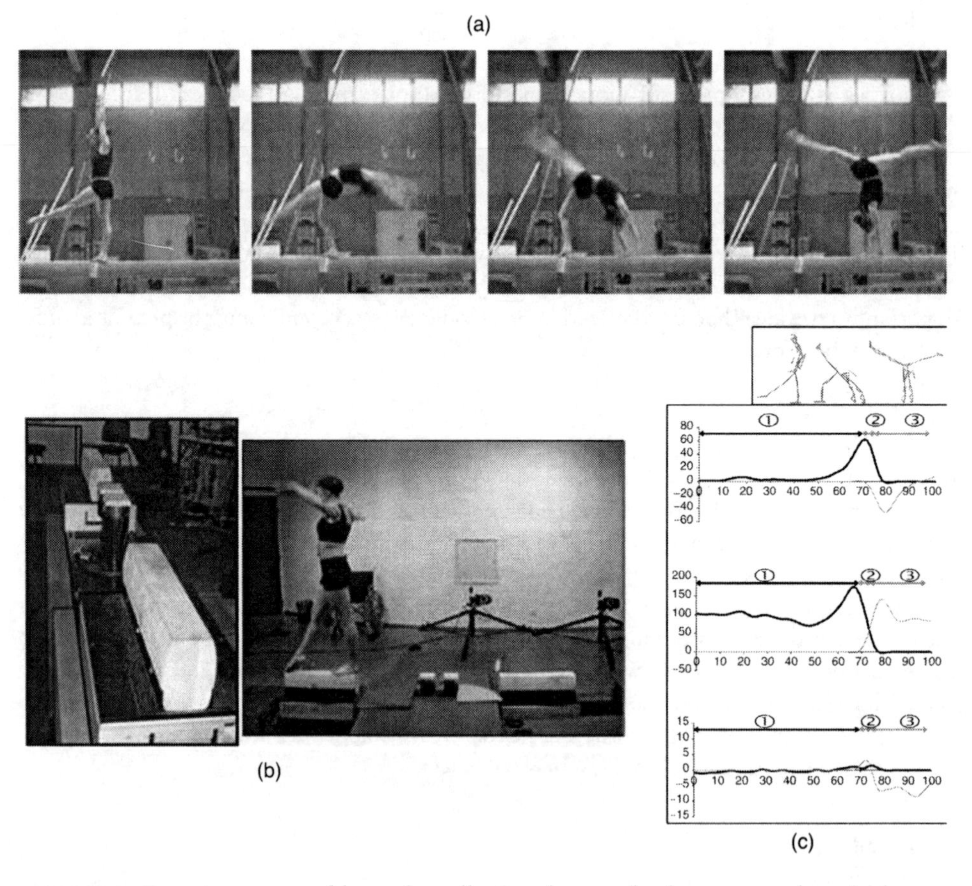


**FIGURE 10.1** Force-instrumented beam for collecting data on the forces exerted on the beam – for example, when performing a back and walk-over handstand (a). Four pieces of beam are fixed on four Logabex force-plates so that they correspond to the foot and hand support target areas. The steel rail supporting the force measuring instruments can be raised to different heights (b). The different components of the reaction forces acting during the foot and hand support phases can be collected for the analysis of the dynamic balance control (c).

weight of a seated person's body is applying a force to the chair – if you put a scale under the legs of the chair you would find that there is a measurable force. However, the chair does not move in spite of the applied force. In the Imperial/English system a force of 1 lb will produce an acceleration of 1 foot per second squared. In the metric system the unit is the newton. A newton is the amount of force required to accelerate 1 kilogram by 1 metre per second squared.

Figure 10.1 shows force-measuring instruments that are required for collecting the forces exerted on the beam during the foot- and hand-support phases when performing a back walk-over to handstand. Force transducers or force platforms convert the magnitude of force to an electric signal. Nowadays, these instruments are connected to computers which collect, process, plot and analyse the recorded forces.

## 10.4 Internal and external forces

Gymnasts also need to consider that they can produce forces from muscle tension (i.e., internal forces) and these forces can be seen in accelerating limbs and body. The body of the gymnast can also experience external forces, primarily gravity, and also the elastic forces of the apparatus, such as springs in the spring floor and the bend and recoil of a rail or bar. For example, Figure 10.2 shows the flexion of the springboard during the takeoff executed by a gymnast to perform a vault. When the deformed springboard is recoiling, the recoil generates an external elastic force that moves the gymnast over the vault table. The recoil ability of the springboard depends on its coefficient of elasticity, which is mainly determined by the nature of the materials from which the apparatus is made. The greater the degree of the springboard recoil, the greater its coefficient of elasticity.

## 10.5 Newton's laws of motion

Isaac Newton codified motion into laws. Newton's laws apply nicely to the world as we see it. However, Newton's laws have been theorized to break down when you consider the very large distances of outer space, very small distances of the subatomic and very fast motion of things moving at the speed of light. Fortunately, gymnastics movements fit nicely into the realm of

**FIGURE 10.2** Elasticity of the springboard. During takeoff the flexure of the springboard is indicated by the trajectory of the markers. The spring properties of the apparatus are an important source of elastic forces used for performing many gymnastic skills.

Newton's laws and these laws have been well understood for hundreds of years. Some mysteries still remain. For example, Newton's first law has never been proven directly because there isn't a situation in which a body has zero force applied to it. There are always forces being applied as a result of planetary rotation, orbital revolution and gravity (Hay, 1973).

### 10.5.1 First law

> A body continues in its state of rest or motion in a straight line unless acted upon by an outside force.

Newton's first law is also called the 'law of inertia' or sometimes the 'laziness law'. Objects at rest tend to stay at rest. Objects in motion tend to stay in motion. In either case, these are examples of an object's 'reluctance' to change motion. An object's inertia is proportional to its mass, which is also proportional to its weight. A more massive body is harder to accelerate than a less massive body.

Once a body is in motion it has 'momentum'. Momentum is described as the quantity of motion in a body. Momentum is the product of the object's mass and its velocity. Increasing an object's mass and/or velocity increases the object's momentum and decreasing either term decreases the momentum.

### 10.5.2 Second law

> The rate of change of momentum of a body is proportional to the force applied to the body, is inversely proportional to the body's mass, and acts in the direction of the applied force.

A change in momentum per unit of time is shown in the following:

$$\textit{force is proportional to } \frac{mv_f - mv_i}{t} \tag{10.1}$$

Where: $m$ = mass
$v$ = velocity
$t$ = time

When the mass of the body doesn't change, the above statement changes to:

$$\textit{force is proportional to } m\frac{v_f - v_i}{t}$$

Of course, the second term (the division), is acceleration, as you should recall from the section on linear kinematics. Multiplying one side of the equation by a constant ($k$) we get:

$$F = kma \tag{10.2}$$

The constant can be removed by noting:

$$1\,newton = k \times 1\,kilogram \times 1$$

And then

$$1 = k \times 1 \times 1$$

Thus, the equation for force becomes the more familiar:

$$F = ma \tag{10.3}$$

From a practical standpoint, a coach or instructor can often see the effects of a force on a gymnast when the coach or instructor is spotting. For example, a gymnast may be performing a tumbling sequence ending with a somersault. The coach assists the gymnast by 'lifting' him/her slightly during the spot and thereby accelerates the gymnast in the direction of the applied force. Every spotter also knows intuitively that an older and probably heavier (more massive) athlete is more difficult to spot because it is more difficult to change the momentum of the heavier athlete.

### *10.5.3 Third law*

> For every force that is exerted by one body on another there is an equal, opposite, and simultaneous force exerted by the second body on the first.

The above is just one of several ways of stating Newton's third law. According to this law, forces always work in pairs. A simple vertical jump can illustrate. As a gymnast performs a jump he/she pushes against the Earth with a force exceeding the gymnast's weight (if he/she leaves the floor and travels upward) and the Earth pushes against the gymnast with an equal, opposite, and simultaneous force. Of course, the mass of the gymnast is so utterly tiny compared to the mass of the Earth that the gymnast moves while the Earth's movement is imperceptible.

A word on the 'simultaneous' term in the definition of Newton's third law is merited. Experience has shown that people often misunderstand the idea of this law by confusing aspects of jumps from elastic surfaces. If our vertical jumping gymnast was standing on a trampoline it appears to the eye that he/she pushes downward against the trampoline and then 'waits' for a second or so for the trampoline to push back. The conceptual difference is that there is never a wait for the second force to arrive on the first body. The illusion provided by the trampoline is that some elastic energy is stored in the springs of the trampoline that is later applied or returned to the jumping gymnast. This is a separate pair of forces than those that occurred during the initial gymnast's downward push against the trampoline bed. One of the ways of stating Newton's third law is that for every action there is an equal, opposite, and simultaneous reaction.

However, even in the statement above, there is no special meaning in mechanics for 'action' and 'reaction' and this makes the statement imprecise (Hay, 1973).

Sadly, the phraseology of this fundamental law of motion has been borrowed by other areas of study to demonstrate a conceptual point while ignoring the mathematical relationships. For example, the idea that for every action there is a reaction in a social setting may be partially true, but using Newton's law as justification for popular psychology is inappropriate.

## 10.6 Impulse

Impulse is the product of the force applied and the time in which force is applied. To increase impulse you can increase force, time or both. The unit of impulse is the pound-second in the Imperial/English system and the newton-second in the metric system.

Impulse is an important concept in gymnastics because of the nature of how the principles of mechanics meet biology. Although it makes mechanical sense that you can increase impulse by increasing the time of force application, in gymnastics there is nearly always a limited window of time in which force can be applied in order to effectively perform the task. For example, the time of application of the downward and horizontal forces is limited by the mechanics of performing an effective takeoff. If the time is lengthened too much then the gymnast rotates about his/her feet to a position that can make the somersault trajectory 'long and low' to a point where the takeoff becomes so dangerous that common sense would deem the increased time on the floor utterly unreasonable. There are also other constraints on impulse from biology such as the limited time in which muscle peak forces can be produced in terms of transmission of forces to bone, energy supply, neural recruitment, strength, pennation of muscle fibres, angles of joints relative to ideal lines of pull and unrecoverable and unusable transfer of muscle tension force to heat.

### 10.6.1 Impulse–momentum relationship

Impulse is a change in momentum. If we begin with force:

$$F = ma$$

and then substitute the equation for acceleration, we get:

$$F = \frac{m(v_f - v_i)}{t}$$

Then distribute the mass:

$$F = \frac{mv_f - mv_i}{t}$$

rearranging:

$$Ft = mv_f - mv_i \tag{10.4}$$

A first example of this principle at work in gymnastics can be seen by gymnasts performing a standing back somersault. A typical performance error, usually caused by fear of failing to rotate backward effectively, results in a gymnast 'pulling' his/her feet off the ground quickly and 'early', reducing the time of force application. Based on Equation 10.4, the incomplete thrust of the legs also shows that the change in momentum will be suboptimal, which usually causes a low flight trajectory for the somersault. A second example illustrates how different arm-swing techniques produce appropriate takeoff impulse for performing a forward somersault on the floor (Figures 10.3 and 10.4).

Figure 10.3 shows that the gymnast's approach velocity produces at the beginning of the takeoff an initial momentum which is the result of great forward and downward velocity components. This existing momentum must be changed so that an appropriately directed momentum at the end of takeoff could be produced for performing an adequate flight trajectory during the aerial phase of the somersault. Thus, it is important to take into account the direction of the new momentum that was redirected. This last idea requires a decrease in the forward momentum and an increase in the upward directed momentum during a very short period of time – the impulse.

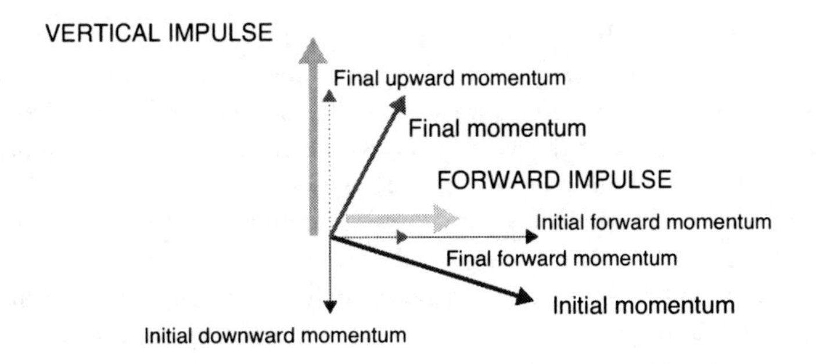


**FIGURE 10.3** Linear impulse and momentum produced during the contact phase on the ground for performing a forward somersault.

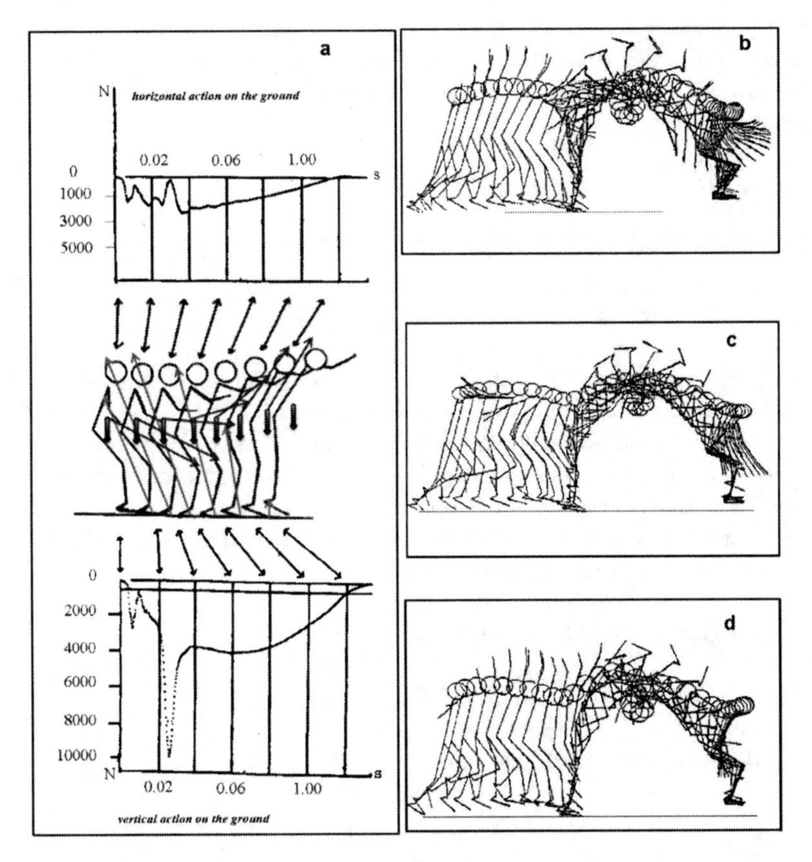


**FIGURE 10.4** External forces exerted during takeoff and change in the body's velocity when performing a forward somersault. The horizontal and vertical components of the ground reaction force (grey vector) and the gymnast's weight (double arrow vector) are plotted according to the stick-figure kinogramm (a). The effect of these two external forces is to produce the required variation of the gymnast centre of gravity velocity (black vector) during the takeoff phase. Three forms of arm swing were commonly executed to produce the appropriate takeoff impulse:

(b) backward/upward arm swing or reverse lift technique
(c) classic forward/upward arm swing
(d) overhead arm throw technique

Figure 10.4(b-c-d) shows that different arm-swing techniques initiated during the support phase before takeoff can produce appropriate acceleration for generating a ground reaction force that is sufficient to perform a forward somersault on the floor. Figure 10.4(a) shows the effect of the external forces exerted during the takeoff on the change in the gymnast's velocity. Performing a forward somersault requires generating high vertical peak force (18 times bodyweight) during a very short time of takeoff (0.10s). During this short period of time, because of the impulse due to the reaction of the ground in response to the takeoff force generated by the gymnast, there is a change in the direction of the body's momentum. Because the 'overhead arm throw' technique produces weaker ground reaction peak force to execute a required takeoff, this technique may be appropriate to prevent gymnasts from ankle ligament sprains.

## 10.7 Work

Given that a force accelerates a body, what do we need to know about the actual motion of the body? Work is defined as the product of the magnitude of the force and the distance the body moves. The force and the displacement of the body are in line with each other, or collinear.

$$W = Fd \tag{10.5}$$

Where: $W$ = work done by the force
$F$ = magnitude of the force
$d$ = relevant distance

When the force acts in the same direction as the motion of the body, work is considered positive. When the force acts in the opposite direction from that in which the body moves, the work is considered negative. When a gymnast jumps to catch the high bar or rings, the work performed to jump upward is considered positive and the influence of gravity (weight of the gymnast) is considered negative work. For example, if the gymnast has a mass of 50 kg, and thus a weight of 490.3 N, and jumps 50 cm (raises his mass 0.5 m) to reach the bar by applying 850 N of force to the mat beneath the bar, then:

$$\textit{work due to gravity} = -(490.3 \times 0.5) = -245.15\ Nm$$

$$\textit{work done by gymnast} = 850 \times 0.5 = 425\ Nm$$

$$\textit{total work} = 425 - 245.15 = 179.85\ Nm$$

## 10.8 Power

Work is a relatively simple concept, with no mention of how much time is required to perform the work. By simply computing the work ($Fd$) completed in a period of time ($t$) we have a new concept – power. Power is the rate of doing work.

$$\textit{power (watts)} = \frac{\textit{work}}{\textit{time}} = \frac{Fd}{t}$$

$$P = \frac{W}{t} \tag{10.6}$$

## 10.9 Conclusion

Linear kinetics involves a variety of concepts that refer to invisible forces and their visible effects on bodies at rest and in motion. A great deal of movement can be explained by kinematics and linear kinetics. Some things that linear kinetics borrows from earlier sections is the idea that a force is a vector quantity, which means that forces can be decomposed to components and combined into resultants.

Forces underlie all motion. A handy rule of thumb for coaches is that if you see an acceleration (the velocity of the body changes), then you can be sure that a force was involved. Moreover, you can be fairly certain that the direction of the acceleration is the result of a single force being applied or the sum of multiple forces that create a resultant in the direction you observe.

# 11

# ANGULAR KINETICS APPLIED TO GYMNASTICS

*William A. Sands*

Angular kinetics is one of the features of gymnastics that best defines the sport.

Gymnastics is not so much about running, stamina, riding a conveyance, throwing something, hitting something, outsmarting a direct opponent or gaining ground on a standardized playing field – gymnastics is about spinning. Gymnasts flip and twist, rotate and revolve and use flight time and the apparatus to perform these skills. Angular kinetics is about the forces and torques, moments of force and inertia, conservation laws, Newton's angular analogues to linear kinetics and other topics.

## 11.1 Eccentric force application

In order to produce angular motion you need a force, but not just any force. A force that rotates something has to be applied eccentrically. By 'eccentric' we don't mean someone with bizarre behaviour but rather the application of force on the body somewhere away from its axis of rotation or its centre of mass.

Figure 11.1 shows how eccentric forces work. Assuming that the three blocks are free to move, let's say sitting on an ice rink or floating free in space (i.e., without friction), then the left-most block will translate and not rotate. The translation is due to the lack of an eccentric force. The force vector shown goes through the centre of mass, resulting in translation only. The central block in Figure 11.1 shows the forces being applied are parallel to each other, in opposite (or non-collinear) directions, and not directed through the centre of mass. The middle block and its forces make up a particular combination of eccentric forces called a 'force couple'. The force couple shown will just rotate, that is, the block will rotate but not translate. The rightmost block shows the same kind of block and only one eccentric force. The rightmost block will also rotate, but probably not precisely around the centre of mass. This third situation results in rotation because of the applied force and the inertia of the upper portion of the block; however, this situation will result in translation as well as rotation.

Figure 11.1 shows how eccentric forces work. However, note that all of the forces are applied in such a way that the line of application of force is perpendicular to the axis of rotation – the centre of mass. What happens if the force(s) is/are applied obliquely?

Figure 11.2 shows a situation similar to that shown in Figure 11.1.

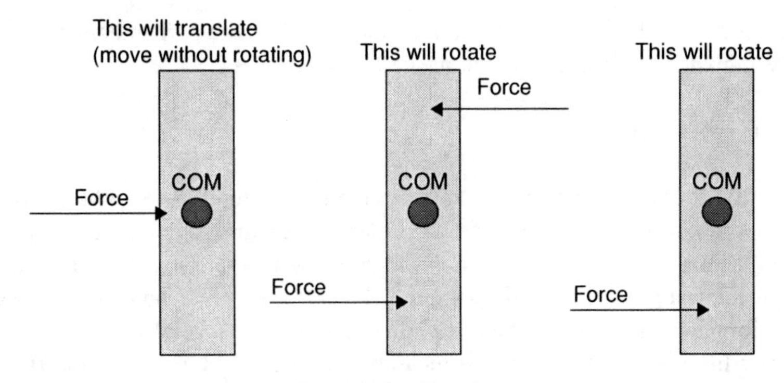


FIGURE 11.1 Translation and eccentric forces.

## 11.2 Torque, moment, force couple

In order to understand Figure 11.3, we must introduce the concept of torque. A torque is not a force. A torque is the measure of the effectiveness of a force, and it consists of two parts. One of the parts is simply the magnitude of the force. A larger force has a greater tendency to rotate something than a smaller force. However, in angular kinetic settings, the placement or line of application of the force also matters. Figure 11.3 shows how forces of equal magnitude (middle diagram) have different levels of effectiveness in turning the object because one is applied closer to the axis of rotation than the other.

The perpendicular distance between the line of application of force (or resistance) and the axis of rotation is called a 'moment'. The 'perpendicular distance' phrase is important – the line of application of a force is not simply where the applied force might be touching the object to be rotated.

The rightmost diagram in Figure 11.2 shows an eccentric force applied obliquely to the object being rotated. More importantly, the perpendicular distance between the line of application of force and the axis of rotation is also shown. When determining the moment of a

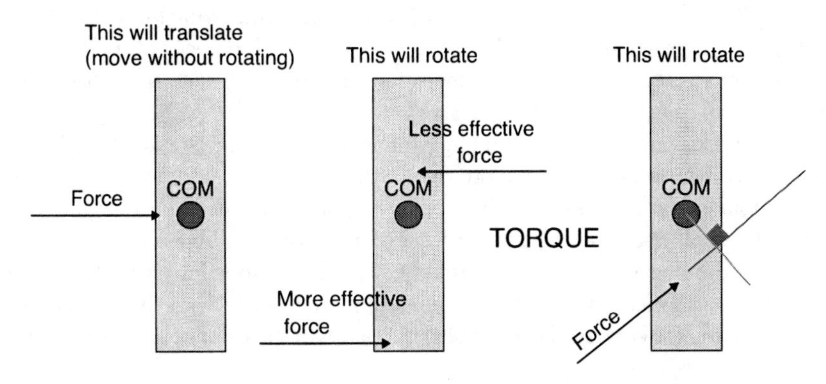


FIGURE 11.2 Moments of force.

torque in the situation shown on the right of Figure 11.2, it is the perpendicular distance that must be determined. The equation for torque can be written as:

$$torque = force \times distance$$

However, this confusing equation looks a lot like the equation we saw before that calculated linear 'work'. Interestingly, you can consider the equations quite similar in concept. Work is the measure of the effectiveness of a force in moving an object by determining the product of the force applied and the distance the object moved. In an angular setting the same idea applies. Torque is the measure of effectiveness of a force in turning an object. The confusing part comes from how the distance is measured. In a linear kinetic setting the distance moved is collinear with the force applied. In an angular kinetics setting, the distance is measured perpendicularly. Thus, to avoid confusion, the equation for torque is sometimes written like the following:

$$T = F \times d^{\perp} \tag{11.1}$$

The symbol $\perp$ indicates that the distance is measured perpendicularly.

An example of the influence of the moment of force or resistance can be seen in a simple door. Note that the hinge (or pivot point) of the door is where the door attaches to the wall. The force applied to open the door is applied to the doorknob. The doorknob is positioned far from the hinge, thus ensuring a large moment of force. If you apply a certain magnitude of force to open the door, the torque to open the door will be the product of the force applied and the perpendicular distance of the line of application of force (usually perpendicular to the door) and the hinge axis. Now let's imagine that the door and magnitude of force remain the same but the doorknob is moved to the centre of the door. You will rapidly find that it is more difficult to swing, rotate, or open the door. Now let's imagine that we put the doorknob a couple of centimetres from the hinge. In this case the inertia of the door and the friction of the hinges may make swinging the door very difficult or impossible to do with the magnitude of force we used before.

In gymnastics we see the same issues arise when we want to create a somersault or salto in the air. For a given force a longer body is harder to rotate than a smaller body. This can be seen in the teaching progression that begins with learning a tuck somersault (knees and hips flexed) through to a pike somersault (hips flexed) to a layout somersault (straight body). We'll come back to this issue later when we consider moment of force and moment of inertia. These are somewhat arbitrary terms for the same thing – a force multiplied by a distance from an axis of rotation.

Both of these ideas are seen in a common gymnastics movement often called a 'trip effect'. A trip effect is shown in Figure 11.3. Note that as the gymnast runs forward and puts his/her feet together for a takeoff, the forward directed force (loosely defined here) will incur a backward directed (equal, opposite and simultaneous) force which will effectively stop the gymnast's feet while the inertia of the rest of his/her body will continue forward. These combined actions and forces will create a force couple that will serve to somersault the gymnast forward. In this case, rotating forward in a layout position to land on the front of the body on the floor, in a pit or on a trampoline.

As the speed of the run, the abruptness of stopping the feet and the distance of the fall increase, the amount of somersault achieved will increase.

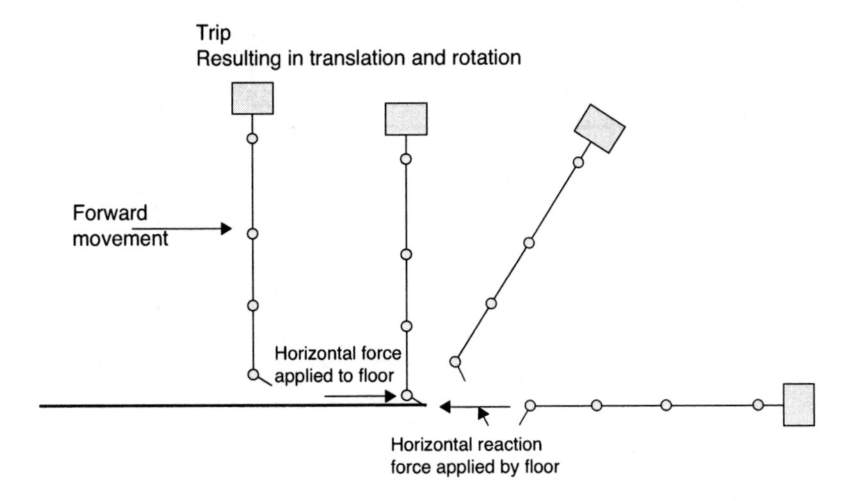


FIGURE 11.3 Trip effect.

## 11.3 Leverage

Levers are of one of the classic simple machines that are used to magnify force or displacement. Levers are described in terms of a force, a resistance (both sometimes arbitrarily named), the distances of each of these from an axis of rotation and an axis of rotation or a pivot point. There are three types or classes of levers, based on where the pivot point is located relative to the force applied and the resistance (Figure 11.4).

The first type of lever is called a type I or first-class lever and is commonly depicted as the teeter-totter or seesaw found in parks where children play. A type II or second-class lever is usually illustrated by a wheelbarrow. A wheelbarrow has the pivot point at the wheel, the resistance is the load between the person lifting the handles and the wheel or pivot point, and the force is supplied by the person lifting the handles. A type III or third-class lever is classically exemplified by most of the joints of the human body, such as the human elbow when the elbow is performing active flexion against some load in the hand.

Figure 11.4 shows the three types of levers with the assumption that all of the levers are at equilibrium or that the products of moments of force and magnitudes of force equal the moments of resistance and the magnitude of the resistance. Note that the type I lever can be made to favour force by moving the pivot point closer to the resistance, as when someone is trying to pry a nail with a hammer or lift a stone with a bar placed on top of another smaller stone close to the stone to be moved. The type II lever will always favour force, owing to the distance of the force from the pivot always being greater than that of the resistance. Type II levers can thereby lift a heavy resistance with a force that is less than the resistance.

Type III levers always favour range of motion due to their placements of the points of application of force and resistance. By favouring range of motion, we mean that for a given displacement of the point of application of force, the displacement of the point of application of resistance is larger. The contrasts in range of motion are shown in Figure 11.5.

Leverage does not apply as much to gymnastics apparatus as to the functional motion of joints of the body. Moreover, classifying joints as particular lever types can be deceptive because load and axis placement may change depending on the specific joint and load

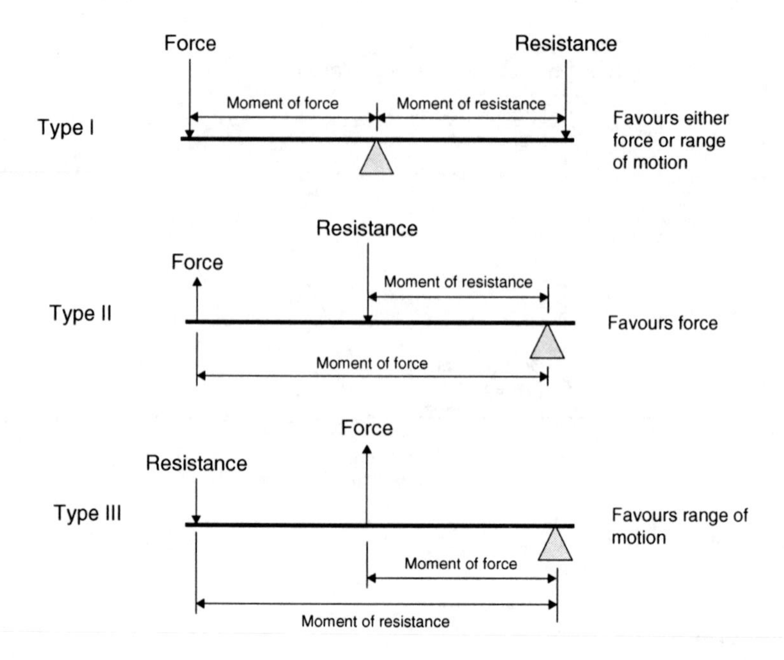


FIGURE 11.4 Lever types and moments.

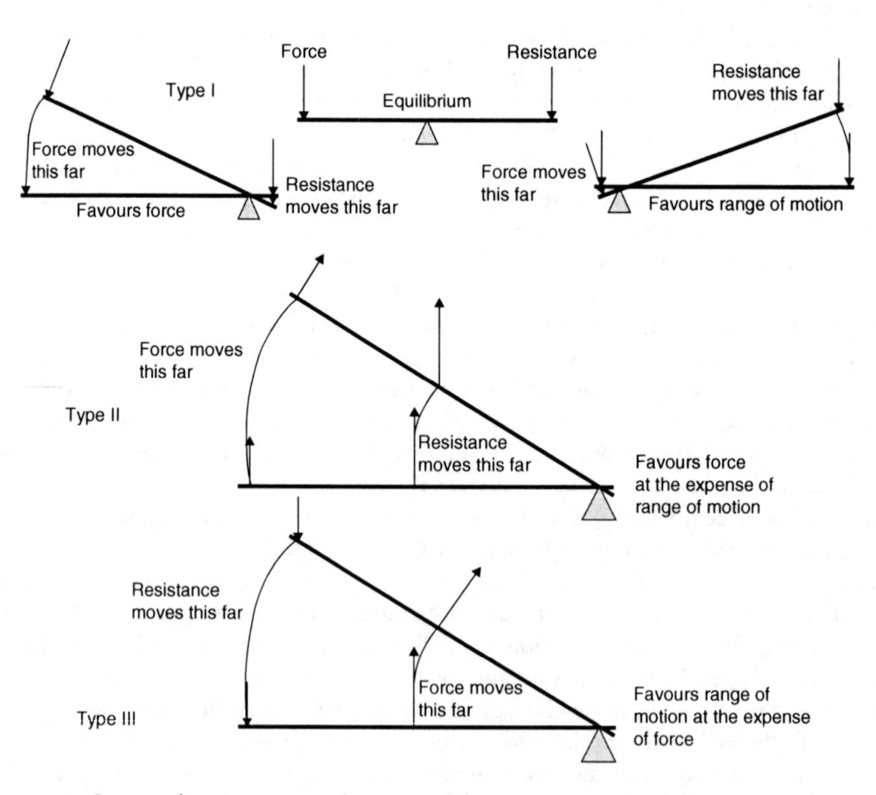


FIGURE 11.5 Levers, showing ranges of motion of the components for each type.

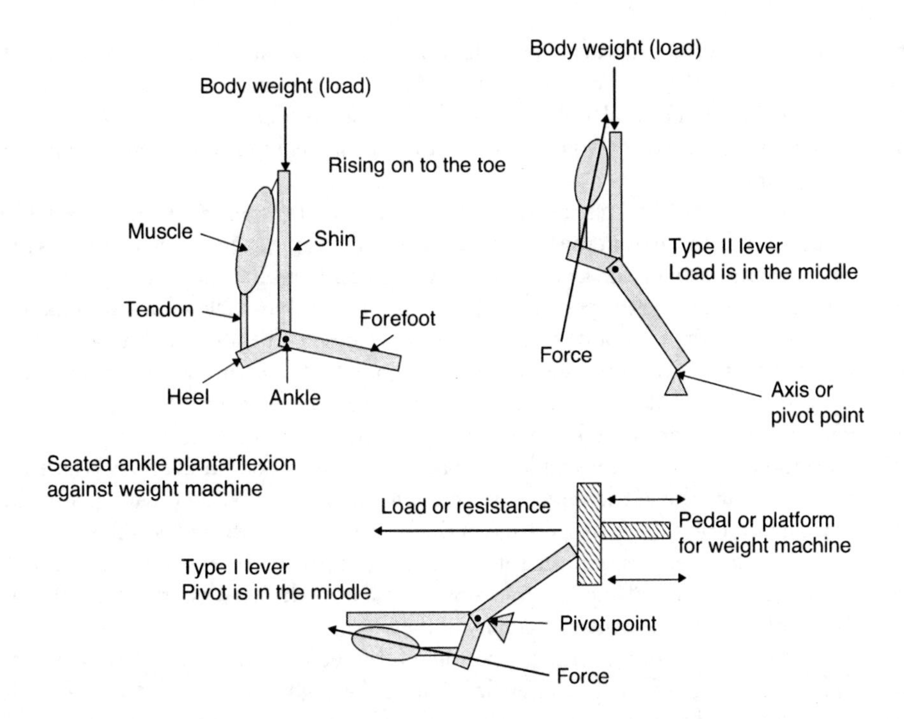


**FIGURE 11.6** Ankle leverage characteristics. Note that rising on the toes exemplifies a type II lever, while pushing with the toes against a leg-press machine to exercise the calf muscles exemplifies a type I lever.

configuration. For example, the ankle joint is shown schematically in Figure 11.6. Note how the variations in lever type depend on the movement performed.

The upper-left illustration shows a schematic of the ankle and the calf muscles that pull on the heel bone. The weight of the body is shown as the load placed on the shin bones and acting through the ankle and foot to the ground. In the upper-right illustration of Figure 11.6 the schematic shows how the load remains acting through the shin and ankle to the foot. The pivot point during the motion of rising on the toes of the foot is accomplished by muscular shortening of the calf muscles, which in turn pulls on the heel bone and raises the load (body weight). The motion described is like that of a wheelbarrow with the load in the middle, the calf muscles pulling upward on the handles of the wheelbarrow and the toes of the foot similar to the wheel of the wheelbarrow.

The lower illustration in Figure 11.6 shows the ankle as it would appear when an athlete is seated at a leg press machine and pushing with his/her toes against the foot platform. By pushing with the feet in the way shown by the lower illustration the load is coming from the leg press machine, the pivot point is at the ankle and the force is the contraction of the calf muscles. Thus, the ankle becomes the pivot point or axis, making the ankle a Type I lever.

## 11.4 Centre of gravity

The phrases 'centre of gravity' and 'centre of mass' have been used throughout preceding discussions without clarification. When a body is acted upon by gravity, we can use the thought experiment of considering that every particle of the body is separately attracted toward the Earth. If we summed the products of all the masses and gravitational attractions of each particle, we would get the total weight of the body. The vectors representing gravitational attractions on all these particles would form a myriad of parallel lines going from each particle directly downward.

However, the particles are not all stacked on top of each other. Each vector would in general be at some distance from the resultant of the summed products of each particle's mass and its gravitational attraction. Determining the resultant of all these particles and their masses requires also knowing the perpendicular distances of each particle from some unknown line of gravity that is the resultant of all these forces. These perpendicular distances are the moments of the particles from the unknown line of gravity. By summing these moments we should achieve a perfect balance of all moments around a particular line of gravity. This new line of gravity originates at the centre of gravity of the body.

For illustration, let's consider a simpler body than the human body. If the body is an empty soda bottle and we lay the bottle on its side, trying to balance the bottle on an outstretched finger, we would find that there is a position of the bottle such that the bottle will balance on the finger. In the balanced position the sum of moments of the particles on either side of the finger will be equal. A plane of gravity projecting upward from the length of the finger through the bottle passes through the centre of gravity of the bottle.

If we then balance the bottle with the bottom on the finger, although more difficult, the plane of gravity projecting from line of the finger upward will go through the centre of gravity. Then to complete the location of the centre of gravity we need to turn the bottle 90 degrees along its long axis and repeat the procedure. Where all three planes intersect the resulting point of intersection is the centre of gravity.

The same thing can be done with a human body, but modern approaches to determining the centre of gravity of the human body rely on computing the moments of each body segment (e.g., thighs, shanks, upper arms, torso, etc.), using measurements or models of the masses of each segment and each segment's centre of gravity to determine the location of the body's centre of gravity. The centres of mass are shown in numerous illustrations in preceding figures.

Knowing the precise location of the centre of mass in the gymnast's body is probably unnecessary. However, practitioners should know that the centre of mass in a standing body lies slightly below the navel and about half way from front to back. Moreover, the centre of mass location changes due to the changing distribution of body segments and the shape of the gymnast's body position.

For example, if the gymnast raises his/her arms above the head then the centre of mass moves upward to reflect the redistribution of mass above the head. If the gymnast starts in a standing position and bends forward to touch the toes, then the centre of mass moves forward and downward relative to the body position. If the gymnast stays in balance over his/her feet, then the centre of mass will always be over the feet (base of support) while lowering. From a standing position, if the gymnast raises one arm up to a horizontal side position, then the centre of mass moves upward and toward the raised arm. Knowing the location of the centre of mass of the body helps determine how the body will move in angular motion settings.

## 11.5 Moment of inertia

The angular equivalent to linear inertia is the moment of inertia. Inertia is a measure of a body's resistance to changes in motion. In an angular setting, the moment of inertia is the body's resistance to rotation. The moment of inertia of a single particle is the product of its mass and the square of its perpendicular distance from the axis of rotation:

$$\textit{moment of inertia} = \textit{mass} \times \textit{radius}^2$$

Since a rotating body has particles or segments in three dimensions around an axis of rotation, the determination of the moment of inertia requires the summing of all the masses and their individual radii about the axis of rotation:

$$I = \sum mr^2 \quad (11.2)$$

Where: $I$ = moment of inertia
$m$ = mass of each particle or segment
$r$ = the radius of rotation of the particle or segment about the axis of rotation

The moment of inertia of a body can be determined in a variety of ways. If the body is some known geometric shape, then its moment of inertia can be easily determined. The same cannot be said for a human body, which consists of multiple, unusually shaped segments that are free to move about. Thus, mass does not change, but the position of the segments does.

## 11.6 Angular momentum

Angular momentum is analogous to linear momentum, with the exception that the mass term is replaced by moment of inertia and velocity is replaced by angular velocity:

$$\textit{angular momentum} = \textit{moment of inertia} \times \textit{angular velocity}$$

$$\textit{angular momentum} = I\omega \quad (11.3)$$

## 11.7 Newton's angular analogues

Like Newton's three linear laws, there are three angular laws.

### *11.7.1 Principle of the conservation of angular momentum*

The first angular law can be summarized as: a rotating body will continue to rotate about an axis with a constant angular momentum unless acted upon by an external force resulting in a torque. The first angular law is also known as the principle of the conservation of angular momentum. This conservation law is of particular importance to gymnasts.

If you consider that angular momentum is fixed at takeoff during a tumbling somersault and the gymnast is free to change shape during the airborne phase, the gymnast can change his/her angular velocity (somersault or twisting velocities) by changing the moment of inertia.

If the moment of inertia is decreased (e.g., move from a layout to a tuck, or from arms outstretched to arms close to the body) then the angular velocity of the somersault, twist or both will increase. Conversely, when the gymnast wants to slow his/her rate of somersaulting or twisting then the gymnast need only increase his/her moment of inertia about the axis of rotation – the centre of mass. The increase in moment of inertia will result in a corresponding decrease in angular velocity. However, since the angular momentum must remain constant or be 'conserved' there is no net change in angular momentum.

### 11.7.2 Angular analogue for Newton's second law

The second angular analogue states that the rate of change of angular momentum of a body is proportional to the applied torque and acts in the direction of the torque. This results in an equation for torque:

$$torque = moment\ of\ inertia \times angular\ acceleration$$

$$T = I\alpha \tag{11.4}$$

The second angular law is shown nicely by a gymnast swinging on the horizontal bar. Figure 11.7 shows an example of how torque and moments interact to cause rotation.

Figure 11.7 shows that the torque of the gymnast in position A is larger than the torque found in position B. Although the gymnast's weight is unchanged in each position, the moment arms show that when the gymnast is horizontal his/her torque is larger and thus the gymnast is more effectively rotated in this position. In order to comply with the second angular law, position A will result in a greater angular velocity during the downswing than position B due to the differences in moment arms.

### 11.7.3 Angular analogue for Newton's third law

The angular analogue for Newton's third law states that for every torque that is exerted by one body on another, there is an equal, opposite and simultaneous torque exerted by the second

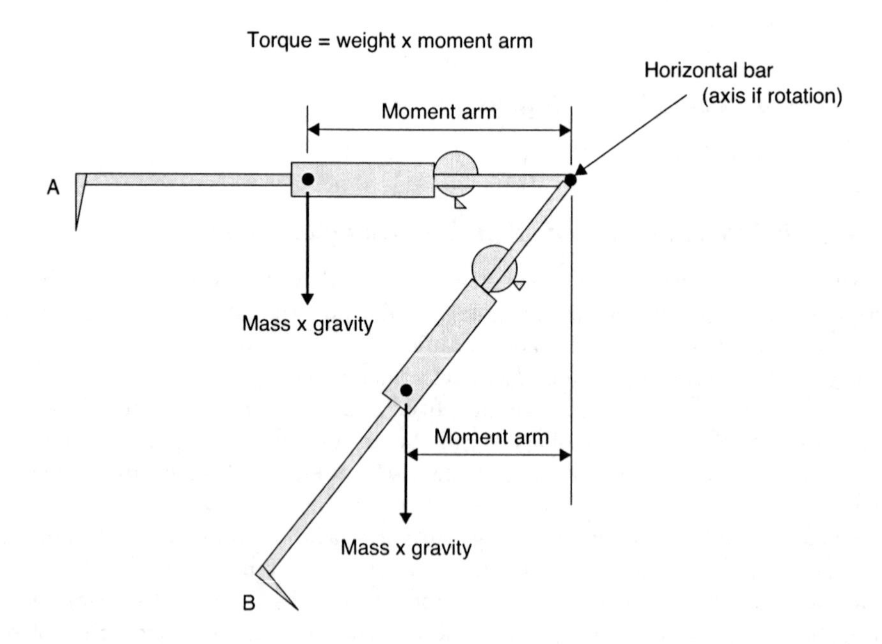


**FIGURE 11.7** Torque comparisons in swinging positions.

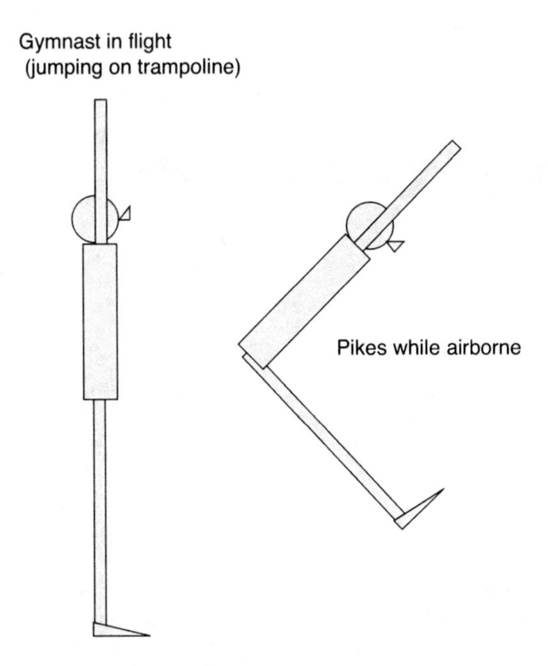


**FIGURE 11.8** Action–reaction in an angular setting.

body on the first. Gymnasts often experience this 'action–reaction' torque problem when they try to change positions in the air. For example, when a gymnast is free in space in a layout position and then pikes, the upper body and lower body will move toward each other.

Unfortunately, the gymnast is often rotating at the time so the illusion that one half of the body is moving more than the other half is exaggerated. This is shown in Figure 11.8.

Gymnasts often use the angular version of action–reaction in order to correct body position while in flight. Qualitative analysis of motion should include knowledge of angular action–reaction because the gymnast who must make large body position changes while airborne usually has serious technique problems. Interestingly, the actual error usually occurs before the visible large body position changes, but knowledge of Newton's angular analogues can serve the practitioner by alerting him/her to problems and where to begin the search for performance (i.e., technique) solutions.

## 11.8 Conclusion

Gymnastics biomechanics is the application of mechanical principles to gymnastics movements. In spite of the style and artistry of gymnastics performances, all gymnasts and their apparatus must obey the laws of physics. Familiarity with these laws and their application can be of considerable assistance in teaching and coaching gymnastics by identifying those motions that represent errors.

# PART II REVIEW QUESTIONS

Q1. A body is measured rotating at 320 degrees per second at the first measurement and 360 degrees per second at the second measurement. The time between measurements is 0.1 second. What is the average angular acceleration of this object? Give your answer in degrees per second per second.

Q2. If an object is projected at 30 degrees from the horizontal at 30 m/s, what are the horizontal and vertical component velocities (to 2 decimal places in metres per second).

Q3. If an object is launched and lands at the same level at 10m/s horizontal component velocity, and has a flight time of 4 seconds, how far does the object go (disregarding air resistance) (Answer in metres)?

Q4. When a figure skater brings her arms close to her body's vertical axis during a spine, he/she spins faster. Explain why does this happen?

Q5. The muscular force of a muscle is 160 pounds, and the muscle is pulling on a bone at an angle of 15 degrees. What are the vertical and horizontal components of this force? (To three decimal places, answer in pounds).

Q6. Draw two stick-figure diagrams of the lower extremity (thigh, shank and foot) and show in Figure 1 a large moment of inertia of the leg during the recovery phase of a running stride. Then draw Figure 2 to show a smaller moment of inertia of the recovery leg than Figure 1. Label the hip, knee and ankle in both diagrams. What could be the implication of the smaller moment of inertia in running?

Q7. What is the distance of a fall that lasted 2.5 seconds (disregard air resistance)? What is the final velocity at impact of this fall? (Answers in metres and metres per second).

Q8. If an object is launched at 35 degrees from the horizontal at 12m/s to land at the same level, what is the peak height of the trajectory and what is the horizontal range?

Q9. If a gymnast has a vertical jump of 0.25 m, what was the vertical velocity at takeoff?

Q10. When a gymnast performs a giant swing, the distance from his hands to his feet is approximately 2.1 metres. If he is swinging at 270 degrees per second at the bottom of the swing, how fast are his feet going – linearly?

Q11. A gymnast (mass = 57 kg) performs a tucked forward somersault on the floor (similar to Figure 9.3). The following table shows the initial and final conditions of the centre of gravity during the takeoff.

| Table | |
|---|---|
| horizontal velocity at touchdown | 4.15 m/s |
| vertical velocity at touchdown | –1.85 m/s |
| horizontal velocity at takeoff | 2.22 m/s |
| vertical velocity at takeoff | 3.01 m/s |
| height at takeoff | 1.12 m |

Use appropriate data/equations to calculate:

- the magnitude of the resultant velocity of the centre of gravity ($v_R$)
- the angle of $v_R$ ($\theta$)
- the horizontal impulse ($p_H$)
- the vertical impulse ($p_V$)
- the peak height of the flight ($H$).

Q12. Explain why the body's moment of inertia is:

- smaller when performing a tucked front somersault than when performing a straight one?
- greater when performing a straight front somersault than when performing a vertical jump with full twist?

Q13. (TRUE/FALSE) By skilfully changing body position in the air (i.e., completely free of support), the gymnast cannot change his/her flight time.

Q14. (T/F) If I push a gymnast (spot) while the gymnast is in flight, and the direction of my push is directly in line with his/her centre of mass, the gymnast will rotate faster.

Q15. (T/F) The centre of mass of a gymnast is a fixed point and does not move during a gymnast's changes in body position.

Q16. (T/F) Torque is same as force.

Q17. (T/F) If the gymnast is somersaulting forward, the vector is drawn from the gymnast's centre of mass toward the gymnast's left.

Q18. (T/F) By increasing the horizontal velocity of the gymnast, when the gymnast stops his/her feet suddenly on the floor, the "trip effect" will be increased and the magnitude of the angular momentum experienced by the gymnast will decrease.

Q19. (T/F) Distance and displacement are the same thing.

Q20. (T/F) If the gymnast's centre of mass is 1.1 metres from the top of the balance beam and she departs the beam from a run at that height and later departs the beam by simply stepping off and dropping straight down, the time of descent in both cases is the same.

Q21. (T/F) The primary determinant of success in vaulting is the horizontal component velocity achieved during the run-up.

Q22. (T/F) When comparing height and distance of a dismount from the horizontal bar or the uneven bars, it is much more difficult to achieve a large range or horizontal distance than a high peak flight of the dismount trajectory.

# PART III

# Psychology and mental training for gymnastics

## Introduction and objectives

*John H. Salmela*

In 1974, I was fortunate as a young rookie professor to have had the opportunity to bring together in a symposium 19 Canadian gymnastic researchers and professors with invited papers, many of whom later became effective national and FIG judges and coaches. We discussed and argued about various issues in aesthetics, motor learning, sport psychology, exercise physiology and biomechanics. Unfortunately at that time most of the reported research was based upon a single study, or upon personal opinions, and in only one chapter were references cited. Still, it was a ground-breaking event and resulted in the publication of *The Advanced Study of Gymnastics* (Salmela, 1976). What, however, has changed since 1976 is that many authors have created substantial bodies of research, have involved themselves in personal interventions with gymnasts at all levels, and have internationally shared their ideas and methods with researchers, coaches, gymnasts and parents.

The objectives of this chapter on sport psychology and mental training are to outline the research on the learnable, and thus teachable, mental skills related to gymnastics, and more importantly to consider how each of the 12 concepts that been developed and published (Durand-Bush *et al.*, 2001) can be applied in actual training and competitive gymnastics. Every theme has an 'implications' section in which the personal experiences from observations, interviews, research and interventions, related to each concept in practice, are developed and then put into practice. Thus the aim of this chapter is to bring together both theory and practice in sport psychology and mental training in a way never previously accomplished for the sport of gymnastics. It is aimed at educated undergraduate or graduate students, gymnasts or coaches.

Consideration will also be given to the specific task demands of gymnastics, the various stages of learning throughout a gymnast's career from different authors' perspectives, ages and genders, and the roles of parents, coaches and mental trainers during these various learning and performing processes.

# 12

# TASK DEMANDS AND CAREER TRANSITIONS IN GYMNASTICS

## From novices to experts, and the stages of learning throughout a career

*John H. Salmela*

### 12.1 Performance task demands in gymnastics

If you were asked to establish a physical and mental training programme for shot-putters, the task would be relatively straightforward. They must propel a 16-lb iron ball as far as possible after having shifted themselves rapidly across a seven-foot circle. Biomechanically, you would suggest that they start from a low position with the shot on their neck and in a position opposite to the intended direction of the throw. They would thrust themselves in a low backward position, and then explosively, with the transfer of energy from their legs to the twisting torso and finally to the arms and fingers, to finally propel the shot during the release. Physiologically speaking, you would train strength and speed of the legs, developing torque speed and arm power prior to the release. Psychologically, you would train the athlete to get pumped up emotionally and to keep in mind self-talk words such as 'drive' and 'explode'. It is pretty straightforward stuff, right?

However, in gymnastics there are six men's and four women's events, each with specific strength, power, balance and skill dimensions, requiring somersaulting, twisting and remaining upside-down for many of the events. Also, the complex dismounts from the horizontal, parallel and uneven bars have inherent elements of danger and fear. In addition, some events, such as the pommel horse, require many rapidly performed movements, demanding fast hand changes and balance, while the rings are performed at a more leisurely pace, but require great power and strength.

Thus, Salmela (1976) and his students did a time-motion analysis of the 1972 Munich Olympics for the final of the men's routines, as well as a classification of the various performed skills. The results for the men were obviously more complex than the analysis of the shot-putter's task. (Figures 12.1 and 12.2.)

The most striking dimension of this data was the diversity of the number of elements and the relative importance of each of these for each event. For example, for floor exercises the women performed 46 elements, although some were small and balletic, while the men performed only 31.2. On the pommel, or side horse, the gymnasts executed a total of 29 movements, while there were only 13.4 on the rings, and, in the same light, on the pommels there were 47 manual releases and, obviously, only one on the dismount from the rings. Given the evolution of the sport over the decades since Munich, it is obvious that the frequencies and

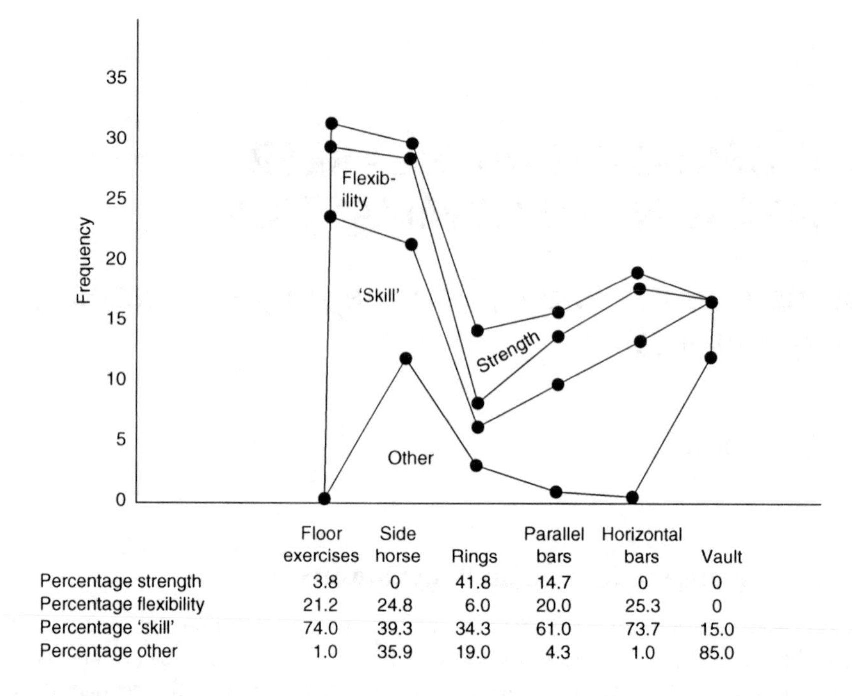


| | Floor exercises | Side horse | Rings | Parallel bars | Horizontal bars | Vault |
|---|---|---|---|---|---|---|
| Percentage strength | 3.8 | 0 | 41.8 | 14.7 | 0 | 0 |
| Percentage flexibility | 21.2 | 24.8 | 6.0 | 20.0 | 25.3 | 0 |
| Percentage 'skill' | 74.0 | 39.3 | 34.3 | 61.0 | 73.7 | 15.0 |
| Percentage other | 1.0 | 35.9 | 19.0 | 4.3 | 1.0 | 85.0 |

**FIGURE 12.1** Frequencies and types of gymnastic components for the men's gymnastic finalists at the Munich Olympics (Salmela, 1976)

numbers of all of these measures would be greatly increased. In recently reviewing the films from Munich, it appeared to me that the men were performing in slow motion, especially on the horizontal bar, as compared to today's very dynamic performances.

If the same analyses were carried out in rhythmic gymnastics (RG), it seems that the patterns would be similar to those of the women's floor exercises, with a greater emphasis on changes in visual orientations and flexibility. RG is highly skilful, and arguably more intricate than artistic gymnastics, since the gymnasts must also perform an open skill task of catching an object. They perform their rolls and twists, which must be timed with the hoop, ball, clubs and ribbon, with no tumbling, but with demonstrations of extreme flexibility.

From a mental training perspective, it is evident that the tasks for a sport psychologist or mental trainer would vary between events. Emotionally, the gymnasts performing on the pommel horse or balance beam would require an alert but more relaxed mental state, while on the floor and vault, for both sexes, the gymnasts must be firing on all cylinders, in that vaulting and tumbling require maximum speed to accomplish the somersaulting and twisting movements.

## 12.2 Implications of gymnastic task demands for learning and performing in gymnastics

It is essential to understand the nature of the tasks which must be trained before beginning to coach, either in artistic gymnastics or RG. In men's gymnastics, the floor exercises and the vault have some common properties, such as the powerful running and tumbling elements;

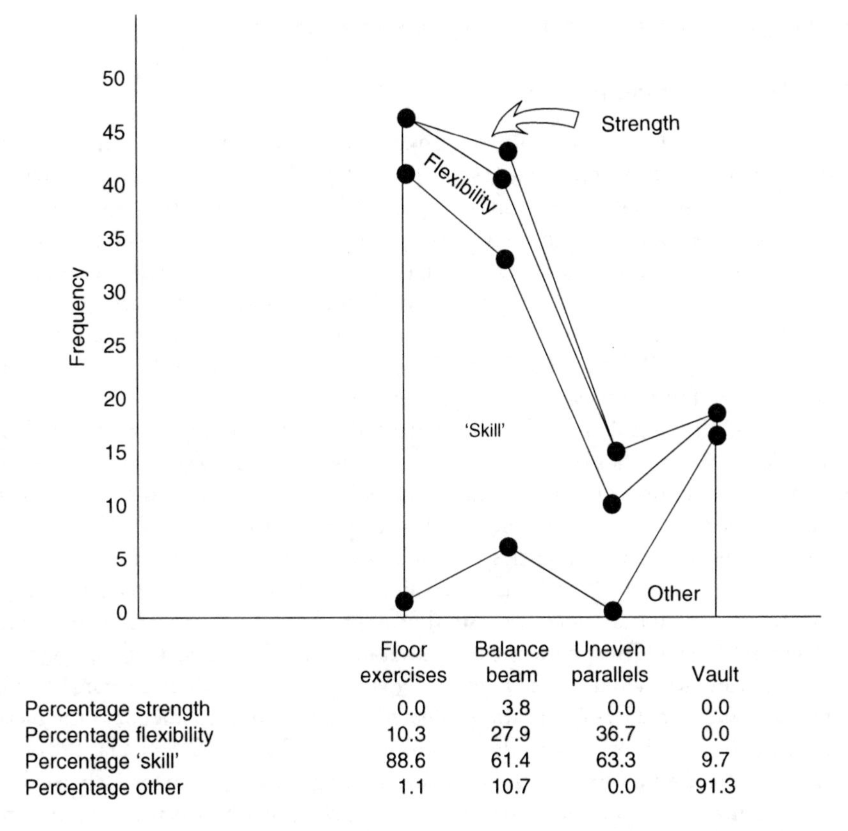


| | Floor exercises | Balance beam | Uneven parallels | Vault |
|---|---|---|---|---|
| Percentage strength | 0.0 | 3.8 | 0.0 | 0.0 |
| Percentage flexibility | 10.3 | 27.9 | 36.7 | 0.0 |
| Percentage 'skill' | 88.6 | 61.4 | 63.3 | 9.7 |
| Percentage other | 1.1 | 10.7 | 0.0 | 91.3 |

**FIGURE 12.2** Frequencies and types of gymnastic components for the women's gymnastic finalists at the Munich Olympics (Salmela, 1976)

however, the pommel horse and the rings have nothing in common, while the horizontal and parallel bars have some similar swing elements.

In women's artistic gymnastics, the floor exercises, vault and balance beam have some identical elements, although the tolerance for errors is less on the beam, which is only 10 cm wide. For women, learning a proper round-off back handspring covers many elements on the above three events. The uneven parallel bars are quite similar to the men's horizontal bar, and for this reason men will often coach this event, as well as the floor and vault, where they are stronger for spotting the performers.

In RG, there are extensive spinning and rolling movements which must be executed in relation to a thrown set of clubs, ribbons, balls and hoops. I would imagine that one of the primary task demands is to be able to launch an instrument into the air in a predictable manner.

Gymnastics technique is continuously changing and, as I have seen while giving mental training courses for the FIG, gymnastics from a technical viewpoint does not at all resemble what I was taught more than 50 years ago. In fact, the great British gymnastics champion Nik Stuart published a book in the 1960s at the end of which he outlined a series of men's routines that he dreamed of for the future. Now almost every element that he envisaged can be realized by a 12-year-old boy!

## 12.3 Stages of learning across a gymnast's career

### *12.3.1 The fixed abilities view*

There are at least two ways of assessing the career structure of developing gymnasts. Ogilvie and Tutko (1966) were probably the first sport psychologists to study 'problem athletes', using paper and pencil tests. But to understand a population of athletes involves first assessing all gymnasts cross-sectionally within each age group, using a variety of valid sport science measures, which I did as the research chairman of the Canadian Gymnastics Federation from 1976 to 1985. The other method is to continuously and personally intervene with them as young gymnasts until their adult career, using 'softer' methods, such as mental training, which I also did from 1985 to 1995 (Salmela, 1989). Both approaches have their strengths and weaknesses, and in many ways are complementary.

In the 1970s, all Canadian sports were obliged to follow the highly successful sporting model of talent identification of the East Germans, who won an extraordinary number of medals for a small country of only 17 million inhabitants. Thus the Canadian Gymnastics Federation, under my guidance as research chairman, created the Testing for National Talent (TNT) programme for men's gymnastics. I consulted with sport scientists who had carried out research in gymnastics, as well as with national and international coaches in Canada, to come up with a battery of multidisciplinary tests that might be related to identifying gymnastic talent. The resulting TNT test battery can be seen in Table 12.1. A portable testing package was constructed and was administered in every Canadian province, where each male competitive gymnast from 10 to 25 years old was evaluated, the data were collated and the measures were correlated with their provincial championship scores (Régnier & Salmela, 1987).

The results were astounding with these 236 evaluated gymnasts, since for each age division the relative contributions of the physical, organic, perceptual and psychological families, which included all variables, changed drastically across the six age groups, with the perceptual and morphological variables usually being dominant (Figure 12.3). It is clear that weight, flexibility, power and all of the perceptual variables are trainable, while limb lengths, height and most selected personality psychological traits are not. What was of interest was that the fixed psychological variables almost always explained the smallest amount of performance variance. Years later, it became clear to me that, with the exception of the pain tolerance variable, all of the psychological evaluations were assessed by paper and pencil instruments which measured fixed personality abilities and did not take into account the learning across the gymnasts' careers.

**TABLE 12.1** List of potential determinants of gymnastic performance in the TNT test

| *Morphological* | *Organic* | *Perceptual* | *Psychological* |
|---|---|---|---|
| Skin folds (4) | Flexibility (19) | Coordination | Anxiety (2) |
| Breadths (2) | Strength | Kinaesthetic sense (2) | Pain tolerance (2) |
| Girths (6) | Shoulder power (6) | Rotation sense | Neuroticism–stability |
| Lengths | Leg power (3) | Foot balance (2) | Extroversion–introversion |
| Height | Speed | Hand balance | |
| Weight | | Time estimation | |
| Morphology (3) | | | |

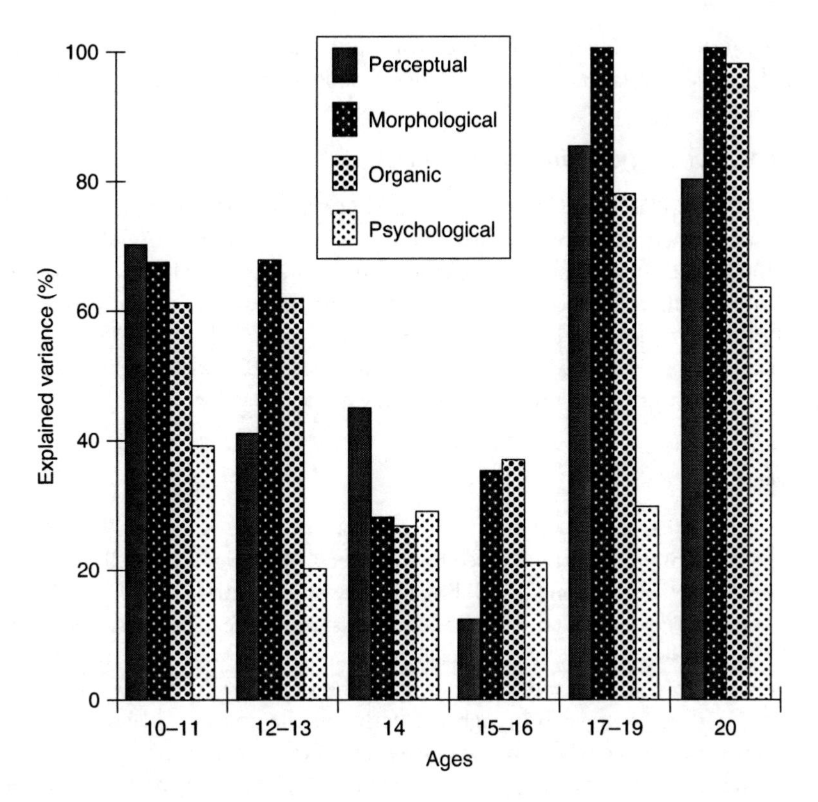


**FIGURE 12.3** Explained performance variance using all of the possible variables in each family of determinants (Régnier & Salmela, 1987)

However, when the number of tests was reduced to the two best predictors, the perceptual variables, requiring balance and spatial orientation, were predominant, but only in the 14–16 age groups did the psychological tests emerge by explaining the most performance variance. Also the morphological and organic variables of height, strength, power and speed were consistently good predictors, but to varying degrees in each age range. Based upon these data, it would appear that psychology, as measured by personality tests at least, was not important in learning, performing and winning, which is contrary to what many gymnasts and coaches report regarding psychology in general (Cogan & Vidmar, 2000). This also confirmed Singer's (1988) view on the usefulness of paper and pencil psychological evaluations.

The fixed-abilities theory's implications for learning and performing in gymnastics are great. For example, body height cannot be trained, and most gymnasts have a short physical stature! According to Ericsson *et al.* (1993), height is one of the few physical characteristics that cannot be trained, even with extensive practice with the support of expert coaches and nurturing parents. It is for this reason that the perceptual variables remained very substantial contributors, especially over the last two age groups (Figure 12.2), since balance and spatial orientating are malleable and continue to contribute to the increasingly complex elements of somersaulting, twisting and balance. Perhaps there is a better alternative of explaining progress in gymnastics performance within a learning perspective. These task-dependent differences (Figure 12.1) are especially true when considering the various task demands of the more stable, strength-related hold positions on the rings, as compared to the highly complex movements, rapid hand changes and dynamic balance that are required on the pommel horse.

### *12.3.2 The learning and intervention views*

Over the last 25–30 years, a number of researchers have traced the developmental milestones, causes and characteristics of expertise development across the careers not only of athletes but of scientists and musicians. What is most encouraging in the results, which used various methodologies and had different objectives, is the consistency of the findings and the progressive refinement of their models. Each of the four outlined studies of Bloom (1985), Ericsson *et al.* (1993), Côté *et al.* (2003), and Durand-Bush & Salmela (2002) added sometimes small but significant elements to Bloom's initial research on expertise development.

#### *Bloom's view*

The nature–nurture issue of whether athletes are born or made is currently tending to side with the environmental perspectives based upon the ecological research of Bloom (1985) and the cognitive research of Ericsson (2007). Bloom wrote an influential book entitled *Developing Talent in Young People*, in which 120 world experts in science, the arts and sport were interviewed, as well as their parents and teachers or coaches. It was shown how the delicate interplay between the athletes and their parents and coaches evolved across three phases of their careers. As they progressed through the early, middle and late stages, their attitudes, activity patterns and goals evolved in a somewhat predictable manner.

In addition, Bloom and his colleagues pointed out that the types of coaching received often changed across the various stages of the performers' careers. Caring, pupil-centred interventions were used early on in tennis and swimming, to increase the students' intrinsic interest in training, and to nurture them to love practising their sport activity, until they were 'hooked' and decided to progress to the next level. The middle phase was more performance-driven, and demanded of the athlete more work and less play, and was supervised by a more disciplined, task-centred coach. During the late stage, sport performance perfection of the skill was required, but this was done in a more collegial manner between the athlete and the coach than in the earlier stages. Bloom also discussed the roles of the parents during the first two stages, where they were required to make family sacrifices regarding, for example, meal times and transportation of the athletes to practice.

#### *Ericsson, Krampe and Tesch-Römer's views*

The acquisition of exceptional performance by athletes has paralleled the development of other expert performers. Within this developmental framework, Ericsson has shown that expert musicians, athletes and other successful performers carried out significantly more deliberate or effortful practice with the intent of specifically improving current performance levels. The fact that exceptional performers did not excel in laboratory tasks which supposedly measured innate capacities led to the conclusion that exceptional performances were driven by environmental factors, rather than by biological or genetic ones, as was the case in the TNT gymnastic project. The main constraints which limited deliberate practice were those of effort, motivation and the availability of human and physical resources. In addition, a metric for the minimum amount of deliberate practice was calculated as at least 10,000 hours, or 20 hours per week, normally spread over a minimum of a 10-year period. Young and Salmela (2002) evaluated Canadian middle-distance runners using their daily running logs and found, contrary to Ericsson *et al.*'s finding, that the effortful training activities were also enjoyable. Ericsson (2007) and his colleagues then transformed Bloom's useful qualitative framework by using quantitative methods, which detailed the practice patterns, the number of repetitions and the practice duration of expert performers, and developed a metric which is applicable in various domains. The necessary required effort and enjoyment levels were also quantified, as well as the presence of physical and personal resources, such as practice facilities and teachers or coaches, to help the performers endure and profit from the necessary but gruelling quantities of practice.

### *Côté, Baker and Abernethy's views*

Côté, Baker and Abernethy (2003) again extended the work of Bloom (1985) and Ericsson *et al.* (1993) regarding the early developmental steps for attaining expertise in sport. This is particularly applicable in gymnastics, and especially for girls, since it is essential that training begins at a younger age than for the boys, since it is easier to provide spotting when they are lighter. They were able to identify and label three developmental stages, based upon interviews with various athletes, parents and coaches. The earliest stages for the sports of rowing and tennis were designated as the *sampling years* (ages 6–12), but could begin as young as five years old. The *specializing years* (ages 13–15) might include 20 to 30 hours of training. This could increase up to 40 hours a week during the *investment years in gymnastics* (aged 16 years and older) (Cogan, 2006). Côté *et al.* (2003), working only in sport, further developed the previous research findings of Bloom and Ericsson, and developed and labeled age-specific guidelines for the stages that were previously outlined.

## The sampling years

Côté *et al.* (2003) showed that successful Canadian and Australian athletes evolved through an initial period termed the sampling years, where children experimented with a number of sports, some of which could be related to gymnastics, such as diving or trampoline. Where Côté *et al.*'s research differed from earlier research was that most of these athletes did not specialize early in one sport, as was found by Bloom (1985) in tennis and swimming, nor by Ericsson (2007) in music, but they experimented with various sports and physical activities. Thus the total number of hours practised in one sport did not attain the magical criterion of 10,000 hours. In addition, these individuals did not all grow up in 'athletic greenhouses', as was the case in many former and present socialist countries; they lived normal social lives, and most went on to finish their university careers.

In more governmentally controlled societies, the early interventions were almost entirely monitored by the coach, whose mind-set was not directed towards the exciting experiences of practising gymnastics – being upside-down, spinning, rolling, hanging and swinging, and being high in the air and then landing safely on a soft and secure gymnastic mat. Rather, many socialist-society coaches begin, early on, to make students practise deliberately and engage in work, rather than play.

## The specializing years

During this period one interesting phenomenon that I have observed when working with the Canadian national age group teams, as they developed into world competitors or Olympians, is what I have called the period of *mental changing of gears*, from being boys to being men gymnasts. When they are from 13 to15 years of age, anything that the coaches or I said as a sport psychologist was digested by the gymnasts as the *truth*! And while they remained at a medium level of performance, this belief in the coach remained fairly constant. Nevertheless, when they then began a six to eight year period of deliberate practice, they learned and perfected the required skills to perform at an international level. However, there was a so far unidentified moment in a skilled gymnast's development during adolescence when this mental changing of gears occurred, and it was real! The parents, coaches and sport psychologist no longer held all of the keys to the universe, in gymnastics nor in their lives, and the students might rebel and disregard advice. Many parents and coaches hate this period, but personally I loved it! They were now autonomous, or, at least, semi-autonomous gymnasts.

### The investment years

Once the gymnasts have committed themselves to a concerted attempt at succeeding at the highest levels, the serious deliberate practice phase begins. in which the intent is to improve on a daily basis, and to refine partially learned skills into highly polished ones. Since the gymnasts are now in their middle to late adolescence, they realize that their investment in the sport requires that they now make choices with the time spent with friends and even in education. The parents are no longer the driving force (nor the taxi drivers!), but the coach now plays a central role in the daily activities of structuring training and practice. As Ericsson (2007) points out, this is not necessarily enjoyable because of the need for the constant repetition of some of the daring gymnastic elements.

#### *Durand-Bush and Salmela's views*

Durand-Bush and Salmela (2002) added another dimension to the above three perspectives with qualitative approaches similar to those of Bloom (1985) and Côté *et al.* (2003). The main difference was that all of the athletes were part of a unique population, since they were either world or Olympic champions, who had repeated their gold medal achievements at these major championships in different years.

#### *The maintaining years*

Following on from the previously cited research on the career stages of experts in sport, the present study extended the three career phases to a fourth stage (Figure 12.4). Four men and six women who had won gold medals in at least two separate Olympics or World Championships were interviewed, and the term the *maintaining years* was coined. No gymnasts were included in the sample, since most gymnasts retire after their first Olympics and get on with their education or obtaining a job. During these years, the athletes trained less but smarter, and they allocated more time for recuperation, since their bodies were battered from years of training and often numerous injuries.

## *12.3.3 Implications of child development for learning and performing in gymnastics*

In this section, consistent but evolving findings have been outlined across the careers of expert performers. The pioneering work of Bloom (1985) over 25 years ago traced the pathways and provided profound implications for the progression of high level performers.

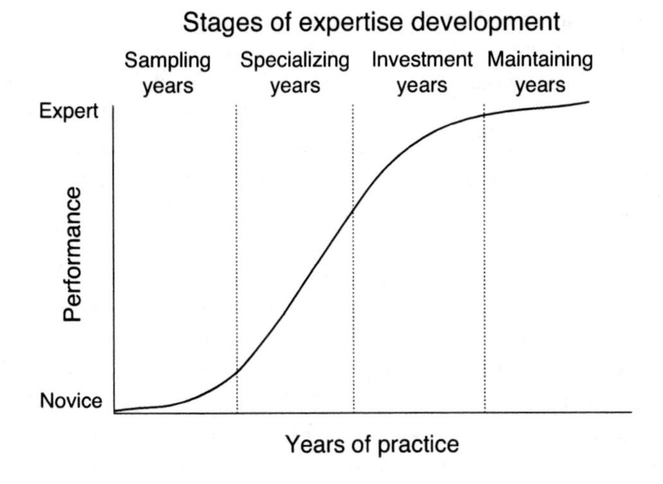


**FIGURE 12.4** Stages of learning and changes in performance across a gymnast's career

But another dimension that is often forgotten is in regard to the practice of gymnastics during free and deliberate play (Côté *et al.*, 2003) during the sampling years. During these early years of a gymnast's career, it is essential that one or both of the parents devote some of their daily routine to becoming a taxi driver, sometimes for two-a-day workouts. Durand-Bush and Salmela (2002) reported that parents of multiple Olympic or World champions rarely had the energy to devote the same amount of time to other siblings and had to find other performance niches in their lives, such as in the arts or music.

Once the gymnasts pass through the specializing or investment years, they still love or respect you, and hopefully both, but now the rules have changed! You as a coach, teacher or parent, do not have the same total control. They now can call some of the shots, or at least participate in the planning of training and competitions, with their coach as a colleague. The extreme stages of deliberate practice may, when viewed from the outside, but appear repetitive and boring, but this is not always the case.

I invited a world-renowned Serbian concert violinist, Dragan Rodosavljevic, to a graduate seminar in Brazil. He remarked that when others watched him repeatedly practise the same piece over many hours, it appeared to them to be boring. He explained that he was curious about minute variations in his techniques and these small changes were not apparent to non-violinists. After 25 years, he was always experimenting with slightly different approaches. It is interesting that the term curiosity is not found in the sport psychology literature regarding adherence to deliberate practice.

Durand-Bush and Salmela (2002) outlined how multiple Olympic and World champions maintained their training with meticulous planning and commitment, but did not require the same quantity for achieving their first gold medal, but improved the quality of training for the second one. They now had more free time and were able to plan for their future employment possibilities when their sport careers ended, since the 10 athletes then were either currently attending high school or university. They reported that imagery, relaxation and self-talk were important skills that they used during this maintaining period and also outlined the importance of their physical and mental recovery to prolonging their careers. There are a number of factors that may have caused the underestimation of how individuals become champions in gymnastics, and the athletes do not always attribute it to the number of hours of deliberate practice but sometimes to the genetically driven notion of *natural talent*.

One of my childhood gymnastic heroes was Abie Grossfield, an American gymnast who competed in the 1956, 1960 and 1964 Olympic Games and was also a long-time coach for the USA men's team. In 1995, I had the opportunity of talking to him at the 1995 Sabae Worlds and I asked why he was such a great gymnast at such a young age, since he made his first Olympic team at 18. He said that he learned all of his gymnastics from 15 to 18 years of age!

Thinking of Ericsson's 10,000 hours of deliberate practice, I asked what he did prior to these three years. He said, 'Nothing – just five years of rope climbing and trampoline twisting and somersaulting'. Now, this is what the Russians are currently doing: massive conditioning at a young age and learning spatial orientation skills on the trampoline. Figure 12.2 reveals the high levels of the contributions of the organic skills (muscular strength and flexibility) and perceptual skills (balance and spatial orientation) during the final stages of gymnastic development. It seems that Abie was wrong in his self-assessment and that the developmental theory of learning was more appropriate!

Duran-Bush and Salmela (2002) reported the remarkable feats of athletes who became multiple Olympic or World champions in separate Games. However, a number of Russian gymnasts like Boris Shaklin and Yuri Titov participated in multiple Olympics and World championships in the 1950s and 1960s for the Soviet Union. What was more impressive was Tanaka's (1987) report which outlined over a 28-year period what he termed the 'Japanese Golden Era of Gymnastics', when 32 Japanese gymnasts competed in multiple Olympics or World contests! Ono and Kenmotsu competed in seven consecutive games, from 1952 to 1964 and 1968 to 1979, respectively, an amazing record of longevity in such a demanding sport.

Two other examples from athletics and one from gymnastics spring to mind regarding passing from the sampling to the investment years and, in the case of gymnastics, remaining in the investment years. The first regards Wilma Rudolph, the three-times Olympic champion in the sprints in athletics (100m, 200m, and 4 x 100m relay) at the 1960 Rome Games; the second was with Lee Evans, world record holder in the 400m and in the 4 x 400m relay at the 1968 Mexico Games. I was fortunate enough to meet them both at a conference in Perugia, Italy, in 1991. While thinking of Bloom's (1985) stages of development I asked them both: 'When did you first think that you were *good* in athletics'? and 'When did you think that you were *great*'?

Wilma was the first to respond and said that she never thought that she would be good at anything in sport, since as a child she experienced no sampling years since she had polio, scarlet fever and pneumonia and for some years lost the use of one leg. But she did say that, as a high school student, she trained at the Tennessee State University with the famed 'Tiger Bells', and was beating their times in practice. These successes in the intervening years transformed her from being good to great in her progress towards the Rome Olympics, where she was great!

Lee Evans was another story, since he was always in good shape, and when he began training as a freshman at San Jose State University, in repeated 400m runs he noted that his more experienced team-mates were lying on the track gasping for air while he was ready for the next series. He said to himself: 'Maybe I do have a future in this sport, and maybe I can become great'. His world record of 43.86 second lasted for almost 20 years.

But perhaps the most amazing gymnastics-related event which occurred during the investment and maintaining years was Dmitri Bilozerchev from the USSR, who won the European championships at the unheard-of age of 16, in 1985. Shortly afterwards, he was involved in a car accident in which he broke one leg in 40 places. But the Russian medical staff were able to reconstruct his leg with pins and straps and two years later after an incredible rehabilitation programme during which he worked on his upper body, in 1987, he won the World gymnastics all-around title! This went beyond the maintaining into the *reconstructive* years! In that year at the Rotterdam World's, his tough mental attitude developed during his recovery, and his performance levels led to his being labelled as *the foreman*, since he always looked in charge of everything.

## 12.4 Conclusion

Gymnastics for men includes six events and four for women. Each event has various and sometimes, extremely different task demands. Some apparatus require balance, others strength and power, and others highly complex and difficult elements. Thus, sport science must be multidisciplinary and specific for each apparatus.

Knowing what are the task-demands for each apparatus: each apparatus has different task demands. The balance beam requires women gymnasts to perform on a beam that is 10 cm wide, and the movements are often equivalent to those performed in floor exercises on a 12 x 12-metre surface. The vault, for both men and women, requires a rapid sprint towards the horse, contact with the horse with their hands at an appropriate angle, somersaulting and /or twisting in the air, 'spotting' the landing point, then 'sticking' the landing with no further movement. The other events require gymnasts to perform movements which are not natural for human beings, including being upside-down, twisting in the air, being supported by only one arm (uneven and horizontal bars), performing movements requiring extreme flexibility (floor and balance beam), dynamic balance (pommel horse) or strength (rings). These elements can be assessed with some sport science methods, but more frequently through the eyes of experienced coaches.

Understand the evolution of the various points of view and methodologies of researchers in gymnastics since 1985. In 1985, Bloom interviewed the best scientists, musicians and athletes in the United States. He found that they followed very similar career paths. During their initial stages of involvement in their activity, they received very caring interventions both by their parents and teachers, and became 'hooked' on the activity. This led them into the middle stage, which required stricter training by the coaches and more parental support. In the final stage, after many years of practice, they worked with the coach in a more collegial way, during which decisions regarding training and performance were decided upon together.

In 1993, Ericsson and his collaborators used a quantitative approach to understand how musicians evolved from novices to experts. Using diaries and questionnaires, they found that expert performers completed at least 10,000 hours of what was termed 'deliberate practice', or practice that was goal-related, but was not necessarily enjoyable, and thus they were able to put a metric on Bloom's work.

Côté, Hay and Abernethy (2003) replicated Bloom's study and labelled the early stage the 'sampling years', when the athletes tried a variety of sport activities; this was followed by the stage in which they would specialize in a given activity for 8 to 10 years, and this was termed the 'specializing years'. This was followed by another period, called the 'investment years', during which the athletes devoted all of their time practising and competing in their selected domain.

Durand-Bush and Salmela (2002) added another dimension to Bloom's model, by interviewing multiple World or Olympic champions, with special attention given to the period between their first and second gold medals, which was termed the 'maintaining years'. Their practice was less intense and more focused, and rest, mental practice, and recuperation were emphasized.

In summary, the models and guidelines first suggested by Bloom provided a qualitative framework for expertise development, which intertwined the career phases of athletes, coaches and parents (Bloom, 1985). Côté and Hay (2002) later labelled the sub-components of the various activities that young, successful athletes were engaged in prior to moving into the deliberate practice regimes during the specializing years. Finally, Durand-Bush and Salmela (2001) considered the development of expertise from the other end of the continuum, when athletes had already achieved Olympic or World Championship gold medals and continued to achieve others over the next cycle, during the maintaining years.

# 13

# COACHING AND PARENTING IN GYMNASTICS

*John H. Salmela*

## 13.1 Coaching

The notion of an autonomous, self-directed athlete's propulsion to world class standards has been put into perspective by the research of Salmela (1996). Expert coaches reported on how they helped shape the learning environments of young athletes by judiciously creating and monitoring achievement goals, and facilitated the acquisition and maintenance of exceptional performance in training and competition by minimizing the constraints which limited their practice, competition and skill development.

Some dimensions that have been so far omitted are the central roles that coaches play, not only in gymnastics, but in other sports as well. Success in gymnastic achievements cannot occur in isolation, nor does it in any other academic pursuits. I spent many hours with great former Soviet Union coaches abroad, such as Leonid Archiaev and Edouard Iarov, and many foreign and native gymnastics coaches living in Canada. Of course, when I compared the best in the world to myself, my shortcomings in understanding the rapidly evolving gymnastics coaching techniques, and the fact that I was now part of the very old school of coaching, quickly became embarrassingly obvious.

Rabelo *et al.* (2001) discovered that young Brazilian soccer players, usually from lower socio-economic classes, received only minimal coaching until they reached the professional ranks as juniors. But they played and practised for an enormous amount of time – at each break in their classes, before and after school, and in nearly all of their leisure time. In contrast, all Brazilian gymnasts received specialized coaching from the beginning of their careers, but during their leisure time did other activities apart from gymnastics (Moraes *et al.*, 2004; Rabelo *et al.*, 2001). This is in agreement with Durand-Bush and Salmela (2000), who reported that Canadian Olympic and World multiple champions spent their youth discovering a variety of other domains and in only a few cases were totally devoted to their selected activities.

Côté and Hay (2002) discussed the importance of two different types of activities in which coaches could facilitate the progression from the deliberate play, sampling years, through the investment and deliberate practice years. Deliberate play involves the child's active participation, is voluntary and pleasurable, provides immediate gratification, and is driven by intrinsic motivation. On the contrary, deliberate practice is not as enjoyable, requires effort, and involves the delayed gratification of rewards (Ericsson, *et al.* 1993).

A coach's decision about the kinds of learning activities to provide and how to decide whether children are capable of profiting from them becomes an important aspect of successful performance. These issues are often discussed, while keeping in mind that a coach's duty, at any level of development, is to stimulate the early fruition of skill and sustain the athletes' motivation for learning and improving. Csikszentmihalyi *et al.* (1993) conducted a four-year study of talented teenagers from the arts, sport and science, who would be in their specializing years, and reported some damning commentaries from the most talented group regarding their teachers/coaches: 'Previous research suggests that teenagers are singularly uninspired by the lives of most adults' (p. 184). This is a strong message for young, aspiring coaches.

However, their most successful mentors were both inspired by the subject matter that they were teaching and reinforced the individual progress and interest of their students. It is not difficult to make this transition to successful gymnastic coaches. Within a positive learning environment, successful gymnasts relished both the hardships and the challenges of training and competition. However, there is a price to pay, since to become great through extensive practice, they have to manage their time better than their normal friends. They were also more often alone than their cohorts, which was not always enjoyable. But, when involved in the activities which they enjoyed, they were more concentrated and happy. However, when these tasks were completed, they reverted more to mindless tasks, such as watching TV or going on the computer, than did their colleagues (Csikszentmihalyi *et al.*, 1993). After interviewing 22 expert, international Canadian team coaches (Salmela, 1996), I realized that my investment in training and coaching graduate students in achieving international academic success, was greater, more intense and better planned than my meagre gymnastics coaching skills.

To date, the psychological differences between men and women or girl gymnasts have not been discussed. One of the first studies on this subject was by Jerome *et al.* (1987) who evaluated a sample of 50 young girls involved in an elite Canadian gymnastics programme. The intention of the study was to determine the factors which differentiated between gymnasts who remained in the programme, and those who dropped out. Twice yearly evaluations were carried out using psychological instruments, as well as others in motor behaviour, physiology and anthropometry.

From my perspective, the most disturbing results were found in the psychological domains. The girls who remained were significantly more obedient and conforming, more introverted, had lower social comprehension, or intelligence, and higher trait anxiety (p. 98). In some ways, these characteristics are similar to those presented by Csikszentmihalyi *et al.* (1993), which indicated that talented teenagers often reported fewer daily positive episodes and psychological states, since they had to sacrifice their social lives, for a period of their adolescence, to achieve excellence. Actually, this profile made these gymnasts extremely coachable, but not every parent would like to have this mental profile for their normal child.

## 13.2 Implications of coaching for learning and performing in gymnastics

Young, talented women are no longer allowed to compete internationally in gymnastics at 13 years of age, but now only at 16. One of my good friends, Dave Arnold, with whom I worked for 20 years, both as a gymnast and with the National Coach, had some interesting insights about coaching young boys and girls. He said that young Canadian girls began training seriously when they were 8–10 years of age, and often skipped the sampling years. He said that the coaches and their assistants imposed intensive conditioning and had them repeatedly practise the same elements with little chance for play and experimentation. The result was that after years of this monotonous training, at the end of their careers, they often totally abandoned the sport, hated their coach, and rarely became either a judge or a gymnastics coach.

Because of the necessity for the men to await post-pubescence to develop sufficient strength and power, the coach still conditioned and trained them, but let them play around on the trampoline, get involved with improvised gymnastic games, and joke with their team-mates and the coach. The coaches' task was to make them love gymnastics and continue to practise into their twenties. Later on, many of these gymnasts went on to become judges and/or coaches at various levels. I am still in contact with the university gymnasts whom I first enjoyably coached in the seventies, and I believe that the fact that we enjoyed each other's presence, liked travelling together and had fun was central. In numerous international competitions, when the Canadian men and the girls were together, the latter often commented upon the fact that the men, with whom I only worked, seemed to be relaxed and happy, compared to how seriously, and almost grimly, the girls viewed their gymnastic training, competition, and life.

## 13.3 Parenting

Bloom (1985) was the first to outline the roles of parents in the development of expert athletes simply by being supportive of their achievement activities and permitting them to choose between practising formally, or to just continue playing. During the middle period, there was more dedication by both the parents and the athletes to continue in their deliberate practice, while during the later stages their support was less physical and emotional, but more financial.

Côté (1999) studied the family environments of elite junior Canadian athletes across their careers. The roles of the parents changed from a leadership role during the sampling years to a follower/supporter role during the investment years. Over all of the years of their child's participation in sport, the parents tried not to pressure them within their given sport. During the investment years, the parents backed off, provided financial support when necessary, but tried not to create additional demands or pressures.

Soberlak (2001) interviewed professional ice hockey players regarding their progress from the sampling through the investment years. Parents did facilitate their *deliberate play*, for example, by building a backyard ice hockey rink, but avoided providing technical instruction. During the investment years, they observed the progress in their sons' performance and provided them with positive, but non-technical feedback. One common thread that characterized all of the studies of Bloom (1985), Soberlak and Côté (2003), Côté *et al.*, (2003) were the transformations of the roles of parents from the sampling to the investment years. During the sampling years the parents had greater direct involvement in their child's sporting activities by teaching and playing with them. This direct involvement, however, decreased as the young athletes moved from the specializing to the investment years, and occurred even less during the maintaining years, when the athletes were young, semi-autonomous adults.

Cogan (2006) convincingly argued that most parents of gymnasts want what is best for their children. But in women's gymnastics, especially, they may exert unnecessarily high expectations on their young athletes, which may have effects that are the reverse of those intended :

> Parents can exert too much pressure if they become overly invested in their child's sport. Sometimes parents can get so absorbed in their own goals and desires for their child that

> they are not aware that the child is losing interest in the sport. Parents can try to live vicariously through their child if they never achieved their sport goals and the child shows potential to excel. In this instance, the children may feel pressure to perform well and keep training when they would rather do something else.
>
> (pp. 646–7)

The practice and play of aspiring soccer players in Brazil was largely unsupervised by parents and often without coaches (Salmela *et al.*, 2004). Rabelo (2002) reported that 78.3% of the families reported that their son's activity in soccer did not change any aspect of their daily life routines, while only 17% reported making adjustments to allow them to participate in sport. This was in contrast to the interviewed parents of, at least, a middle class standing in Canada (Côté & Hay, 2002). However, Moraes *et al.* (2004) found that 24% of middle class and upper middle class parents of tennis players took time to interact with the coaches regarding their children's progress. Finally, Vianna (2002) showed that the tennis parents understandably maintained daily contact with their children at all levels of performance, while 50% of the soccer parents only saw their sons every one to three months at the time of the interviews. Thus the financial status of parents and athletes determined the quality of the interactions between the athletes and parents, and while play, although potentially financially rewarding in the long run in Brazil, their support played a lesser role than putting food on the table and a roof over their heads (Salmela & Moraes, 2004; Rabelo *et al.*, 2001).

## 13.4 Implications of parenting for learning and performing in gymnastics

Social status and parenting in sport are intimately linked. For example, it was shown above that soccer players from poor families from the interior of Brazil received almost no social nor financial support from their parents compared to the middle and upper middle class athletes in private clubs, in gymnastics, swimming and tennis. One poor football player was forbidden to play in a championship final by his parents, but jumped out of the window of his house to play the game anyway; when he returned home after the victory, he was beaten by his parents (Rabelo, 2002)!

I came from a working class family in Montreal and both of my parents worked, and I think that my parents were happy that I was out of the house. They attended only one gymnastics meet, Little League baseball game or Canadian football game. I have known many Montreal gymnasts who made the national team who were poorer than me, and they walked to the gym, without their parents. While travelling in China, Japan, Russia and the Ukraine, I never saw a parent watching a gymnastics training session, so obviously socio-economic elements play both positive and negative roles in gymnastics.

In North America, some girls' gymnastic clubs either keep the parents away with limited possibilities of seeing them train, or place them behind a class enclosure so that they cannot hear the coach/athlete interactions. If not, the parents would complain that their daughter did not receive the same attention or feedback as the other gymnasts, and they then would complain to the club president. So there is such a thing as over-parenting, where the parents become somewhat familiar with the gymnastics terminology and begin advising the coaches on what they should be doing, and exerting unnecessary pressure on their child; or even worse, take a judging course, and to begin judging with little or no practical experience in the sport.

Therefore, parents are essential for the provision of a nurturing environment for the introduction into sport of their children, and then providing them with the necessary resources to practise their sport, which sometimes means driving them to the facilities and to competitions, as well as financing the purchase of equipment and uniforms. Very often, in middle and upper class families, this means rescheduling their daily routines, such as meal times and to accommodate training regimens. Caution, however, should be taken, not to 'over-parent' their children, which could take away from their intrinsic enjoyment of the sport.

## 13.5 Conclusion

In a highly technical sport, such as gymnastics, coaches play a central role for the development of athletes. However, in the developed, non-socialist world, they function in very different ways across the careers of gymnasts. As Bloom (1985) pointed out, during the early years they must be process and athlete centred, otherwise the athletes, who have many life choices, will drop out of the sport. During the specializing years, they must be more demanding, and require that the gymnasts receive and accept strict conditioning, longer hours of practice and confront often dangerous and challenging tasks. Later in the gymnasts' career, during the investment years, they can act more as consultants, since the majority of the training has been accomplished over, at least, ten years of training.

Parents, however, play central roles during the early years of a gymnast's career, since they have to introduce them to the sport, and often have to transport them to the training sites. When they are in the specializing years, they also might have to invest in their travel and equipment, but this often diminishes during the investment years where they are supported by their team or national federations.

# 14

# MENTAL SKILL DEVELOPMENT AND VARIATIONS IN GYMNASTS

*John H. Salmela*

## 14.1 Mental skill learning for enhancing performance

Many approaches have been developed over the years to assess the nature and effects of mental skills for performance enhancement in sport, and to identify those factors that were most critical for sporting excellence. Some of these approaches which could be useful in gymnastics included the following: consulting with mental trainers or sport psychologists, coaches and/or colleagues using existing psychological instruments (Bernier & Fournier, 2007; Mahoney, 1979, 1989; Nideffer, 1988), interviewing of athletes (Orlick & Partington, 1988; Ravizza & Rotella, 1982), and observing gymnasts' behaviours in sport settings (Salmela *et al.*, 1979).

The use of psychological inventories has received much attention by researchers. Fogarty (1995) made a number of useful comments on the administration of psychological instruments within sport settings. He stated that:

> Many of the tests appear to be new, often developed for the purpose of a single study. Furthermore, because most of these new tests are not fully validated, they are not released for commercial publication and consequently do not find their way into the major distribution channels. More importantly, they are not subjected to the formal review processes which most commercial tests have to undergo.
>
> (p.167)

In gymnastics, little sport psychology research has been carried out besides the personality research of Jerome *et al.* (1987), which did not consider levels of expertise, but of dropouts and non-dropouts in an elite club in Canada. The only other study compared the competitive behaviours of top ranked gymnasts, compared to lower classed individuals (Salmela *et al.*, 1980). It seems obvious that the levels of expertise of performers, within any domain, need to be considered, since national and international Canadian athletes in a number of sports, and international Iranian athletes, in all sports, who qualified for the Asian Games, and those who did not, all had different mental skill levels, which narrowed in number as they became more skilled (Salmela *et al.*, 2009). Thus, the understanding of these various sports is important to understanding gymnastics.

## 14.2 How each mental skill affects the others

One thing is clear when assessing, teaching or applying mental skills in sports. First of all, the mental skills should incorporate the three most general conceptual fields in psychology, that of behaviours or actions, cognitions or thinking and psychosomatic elements, or feelings. Secondly, it must be made clear that these general families and their more specific subcomponents are interactive in nature and that changes in one of the components will always affect the other two. For example, if a gymnast receives a knee injury when tumbling (the behaviour), this will immediately affect his/her current emotional state of anger or sadness (psychosomatic conditions), which in turn will affect their thinking – for example, 'they are not worthy and may not make the team' (cognitions).

## 14.3 The development of the OMSAT-3

As the result of research in the field of sport psychology, it has become evident that mental skills play an important role in achieving excellence in sport (Mahoney *et al.*, 1987; Nideffer, 1988; Orlick & Partington, 1988). Orlick and Partington (1988) found that the most successful interviewed Canadian Olympians, who participated in the 1984 Games in Sarajevo and Los Angeles, were seriously committed individuals who believed in themselves and had the determination to accomplish their specific goals. According to Orlick (2008), being committed to excel through the good times and the bad, and believing in one's ability to succeed and reach personal goals are fundamental elements that gymnasts needed to perfect to achieve exceptional performance. Ericsson *et al.* (1993) also demonstrated the vital importance of developing high levels of commitment to overcome effort and motivational constraints associated with daily deliberate practice. Commitment is simply indispensable in the pursuit of expert performance to achieve such a goal.

Furthermore, there have been many attempts to distinguish between those mental skills that are most critical for performance enhancement. Based upon substantial consulting experiences, active interactions with well-known applied mental trainers such as Terry Orlick, Ken Ravizza, Bob Rotella and Len Zaichkowsky, the Ottawa Mental Skills Assessment Tool (OMSAT-3*) was developed over a decade and validated using confirmatory factor analysis, a powerful statistical tool which is rarely used to evaluate the validation of psychological scales in relation to actual sport performance (Durand-Bush *et al.*, 2001).

It should be noted here that the term *sport psychologist* from now on in this text will be replaced by the term *mental trainer*, since the claim to be a psychologist is regulated by law in many countries. But the term *mental trainer* is comparable to *conditioning trainer*, *strength trainer* or *tactical trainer*, and what will be outlined from now on will be related to mental training. Research in many theoretical aspects of sport psychology will be used, but it is believed that mental training relates mainly to the application of mental skills training for sport performance, in a more holistic and interactive manner.

The OMSAT-3 is made up of 12 scales which have been regrouped into the three broader categories of Foundation (*goal-setting, commitment and self-confidence*), Psychosomatic (*stress control, fear control, relaxation, and activation*) and Cognitive skills (*focusing, refocusing, imagery, mental practice, and competition planning*).

Fournier, Calmels, Durand-Bush and Salmela (2002), using the OMSAT-3, found that refocusing skills improved significantly after a 10-month mental training programme, while stress control showed no significant changes. Thus, refocusing appears to be more *state-like*, i.e., capable of being taught and learned, and it is thus a true mental skill. Stress control, however, might be more *trait-like*, in that it showed the lowest scale scores and resisted mental training, and may have been genetically inherent in these elite athletes.

## 14.4 Variations in mental skill patterns over levels of expertise

In the initial development of the OMSAT-3 (Durand-Bush *et al.*, 2001), the instrument was administered to over 200 national and international experienced Canadian athletes (Figure 14.1). The results from the sport psychology literature and common sense were confirmed. Just as would be expected with international athletes as compared to the national level ones, their mental skill levels paralleled their advanced physical skills, and the scores were statistically higher for the international than for the national group on 10 of the 12 mental skills.

The variable of stress control was particularly intriguing given the findings in the other OMSAT-3 studies, which included only high level, international athletes competing in Iran. The significant dimensions which marked the Iranian studies (Salmela *et al.*, 2009) compared to the Durand-Bush *et al.* (2001) study were that all of the Iranian athletes who were assessed were already performing at the international level. However, those who were selected for the Asian Games differed from the non-selected athletes on the OMSAT-3, but only on two scales, i.e., those for stress control and refocusing. These scales significantly differentiated between the two groups and relaxation skill differences approached significant levels. The reduction in the number of scales which differentiated between the groups of international athletes seems logical. As Durand-Bush *et al.* (2001) pointed out, higher scores on the three foundation skills of goal-setting, self-confidence and commitment were central to exceptional performance in all sports. All members of this sample reported high levels on these three

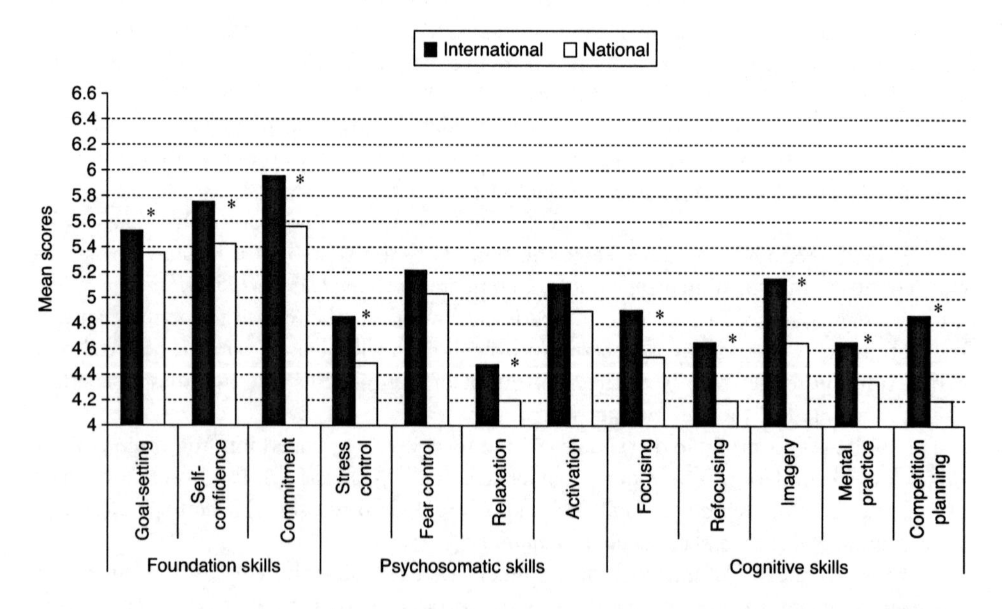


**FIGURE 14.1** Profiles of Canadian national and international level athletes on the OMSAT-3 scales. The '*' represents statistically significant difference ($p < 0.05$) between the groups (Salmela *et al.*, 2009)

foundation scales, with values greater than those reported by Durand-Bush *et al.* (2001) for Canadian national and international athletes (Figure 14.1).

In the second phase, the Iranian medallists were found to be different from the non-medallists on the single scale of stress control (Salmela *et al.*, 2009). This again revealed that there was a consistent hierarchy of mental skill scores across levels of expertise which paralleled their sport performance levels. The present sample of athletes demonstrated the evolution from the 10 OMSAT-3 scale differences reported in the Durand-Bush *et al.* (2001) study, in which only two scales of stress control and refocusing for the selected and non-selected international Iranians, and then only one, stress reactions, differentiated between the medallists and non-medallists. These interactions between expertise and mental skill levels are unprecedented in the mental skills literature (Figure 14.3).

What was of particular interest was the consistency of the findings between the two comparisons, with stress control, acting in a fixed trait-like manner, while refocusing showed significant learning over a 10 month training period, and stress reactions showed no improvements with mental training, and acted in a stable manner (Fournier *et al.*, 2005). What was most striking was the fact that stress reactions and refocusing were consistent among the most highly correlated scales in both the Durand-Bush *et al.* (2001) study with international and national Canadian athletes and the present study for Iranian international athletes, for both the selected and non-selected athletes. The reasons for these associations are as yet unclear, but it is interesting that the levels of stress reactions for both the Fournier *et al.* (2005) and the Salmela *et al.* (2009) studies were also the lowest of all of the OMSAT-3 scale values, and were most resistant to mental training.

One conceptual explanatory hypothesis might be the following: the trait-like nature of the stress reactions and their lowest rankings appeared to be unchanged after mental training. It could be that the elite athletes either persevered to improve this dimension with extended

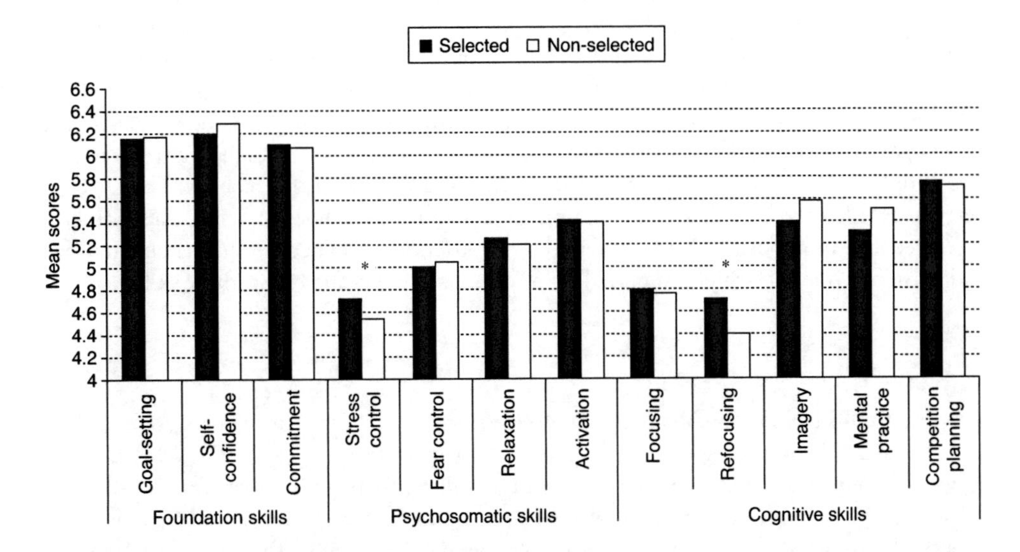


**FIGURE 14.2** Mean OMSAT-3 scores for Iranian athletes selected and non-selected for the 2006 Asian Games. The '*' indicates statistically significant differences ($p < 0.05$) between the groups (Salmela *et al.*, 2009)

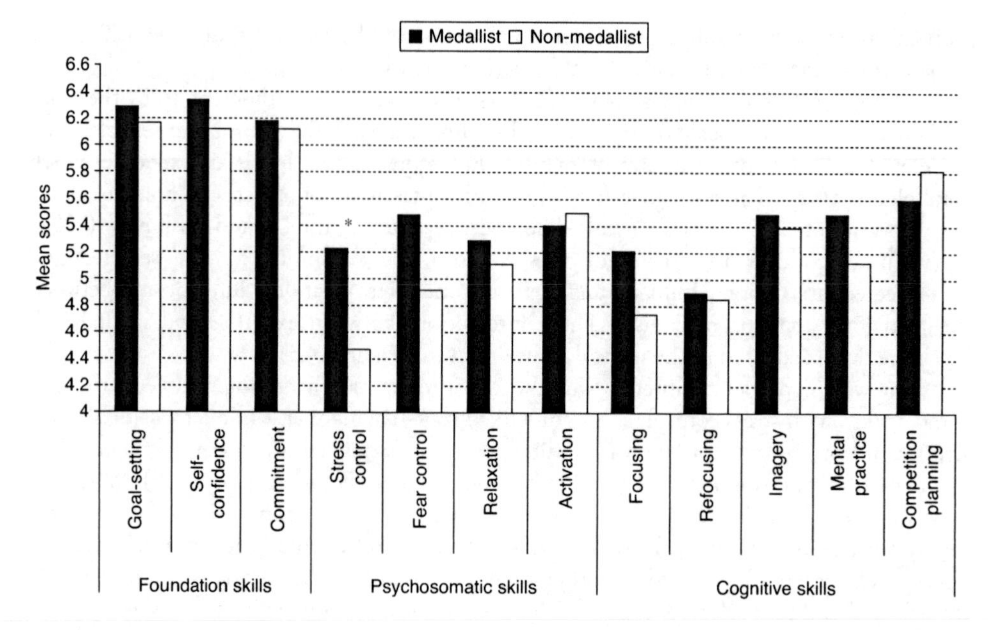


**FIGURE 14.3** Mean OMSAT-3 scores for Iranian medallists and non-medallists in the 2006 Asian Games. The '*' indicates statistically significant differences ($p < 0.05$) between the groups (Salmela *et al.*, 2009)

practice, or were born with greater mental toughness or resistance to stress. The high level of learning of the refocusing skills partially reinforced the persistence of the differences of the stress reactions between the two groups in both Iranian studies. All selected and non-selected athletes did not differ on the three foundation skills of goal-setting, self-confidence and commitment, perhaps due to their high levels of expertise, which also validated their international athletic status, as was predicted in the OMSAT-3 model (Durand-Bush *et al.*, 2001).

Of particular interest was that the five Iranian coaches who had daily contact with these athletes were also asked to rank the importance of the same mental skills of their athletes, based upon their perceptions of the players' strengths and weaknesses. Significant discrepancies were found between the rankings of the athletes and those of the coaches. The coaches perceived athletes to be skilled at maintaining focus and self-confidence, while these athletes perceived just the opposite. Coaches viewed their players as poor at controlling their emotions and tension, while the athletes reported themselves as being proficient in these skills.

## 14.5 Conclusion

Mental skills have been shown to be better predictors of sport performance, compared to fixed, innate psychological parameters, such as personality tests or IQ. The OMSAT-3 was developed to assess a variety of mental skills, which were shown in the sport psychology literature to be essential for expert performance. The essential ingredient was that these skills could be taught and learned by developing athletes and coaches, sometimes with the help of a mental trainer.

The basic elements were the foundation skills, without which expertise in gymnastics, or any other domains in sport, science or music, would be impossible. These three skills included goal-setting, or having high expectations for potential achievements. The second category included psychosomatic or emotional skills, such as stress and fear control, relaxation and activation, some of which would be used to different degrees in training and competition. The third category was cognitive or thinking skills, which included imagery, mental practice, focusing, refocusing, and competition planning.

All of these mental skills were tested in athletes of different levels, and some skills differentiated successful athletes at different performance levels and medallists from non-medallists at international competitions (Durand-Bush *et al.*, 2001; Durand-Bush & Salmela, 2002; Salmela *et al.*, 2009).

# 15

# OMSAT-3 MENTAL SKILLS ASSESSMENT OF GYMNASTS

*John H. Salmela*

## 15.1 Goal-setting

Goal-setting is perceived to be an essential performance enhancement skill by many researchers (Burton, 1993; Harris & Williams, 1993). More precisely, it is believed to help athletes focus their attention, remain intense and persistent, increase their self-confidence, and control anxiety (Burton, 1993). Gould (1998) also suggested that athletes set specific, measurable goals which are difficult, but realistic to achieve, to maximize their effects. Others have revealed that athletes' performances can be additionally increased if they set short- and long-term goals (Harris & Williams, 1993), as well as performance goals, such as increasing their number of successful routines or perfecting problematic elements, as opposed to outcome goals, like defeating a given opponent or obtaining a certain score on an event (Burton, 1993).

'I set difficult but achievable goals' in the goal-setting scale of the OMSAT-3 was an excellent discriminating item between the Canadian national and international athletes, which suggested that elite athletes set goals, but not just any type of goal – rather, challenging goals that they believed they could achieve (Durand-Bush *et al.*, 2001).

### *15.1.1 Implications of goal-setting for learning and performing in gymnastics*

It would be unrealistic to assume that aspiring gymnasts would set high level goals on a daily basis to be proficient in such a multi-faceted and difficult sport as gymnastics. When young girls first saw on television Romania's Nadia Comaneci and the Soviet Union's Nelli Kim scored perfect 10s during the 1976 Montreal Olympics, or Mary Lou Retton who helped the USA win the team title in 1984 in Los Angeles, gymnastic club registrations skyrocketed across the world in women's gymnastics. However, even though goals in sport can maintain one's focus, persistence and self-confidence, this is not always sufficient. But gymnasts for whom this is their only livelihood in financially depressed societies may persevere in their attainment of these lofty goals, since they are dependent upon state support for success and competition cash prizes for a good future life. This is also now the case in Brazil, where successful athletes such as Deanna dos Santos and the Hypolito family can now attract lucrative, financial contracts.

I often ask my students in some FIG courses who would like to be seventh best in the world in anything. Most people raise their hands. I then reply that the men's all-around champion, Li Ning from China, the 1984 Los Angeles Olympic all-around champion, did not share this same goal. I was with the National Coach of Canada with Curtis Hibbert in the warm-up gym preparing for the horizontal bar finals at the 1987 Rotterdam World Championships. At that same time, the medals were being awarded for the pommel horse finals, and shortly afterwards, all eight finalists entered the gym. Li Ning had fallen from the pommel horse and was awarded seventh place. When he walked into the warm-up hall, he had his certificate in hand, he threw it in the air in disgust and it landed close to where we were warming up. Since he was obviously going to leave it there, I went over and kept it as a goal-setting illustration for future classes and courses in gymnastics. Obviously, seventh place was not a goal for Li Ning!

Concerning the lifelong-term goal-setting of each university student, I often ask them in my classes, courses or workshops the following question: 'How many of you would like to be an Olympic or World champion in your favourite sport'? Almost all of their hands immediately shoot up. But when I ask them: 'How many of you are willing to work at your sport for six days a week, for 20 to 30 hours weekly for the next 10 years'? and almost all of the raised hands disappear. Again, goal-setting has to be accompanied by the two other foundation skills of self-confidence, in being able to realize these goals, and their commitment to doing what is necessary in the attainment of these goals, despite the time investment and possible setbacks and normal injuries.

This is what Ericsson (2007) termed deliberate practice, or effortful practice with the intent of improving current performance to the highest levels in the world. But to do so, extremely high goals are required. The normally cited minimum amount of deliberate practice in order to achieve expertise in sport, chess or music is 10,000 hours, although the number of hours with musicians can increase up to 55,000 hours, if they continue during their golden years. During a seminar in Canada with elite gymnastics coaches, I asked them to estimate throughout their careers how many hours they had trained as a gymnast. The answers varied from 3,000 to 7,000 hours for the Canadian gymnasts/coaches, with one exception: a former woman USSR team champion at the Montreal Olympics, Elvira Saadi, who is presently successfully coaching in Canada, reported an astounding 20,000 hours of training!

Personally, I cannot claim more than a bronze medal as a junior at the Canadian national gymnastics championships and a fifth placing in the all-around classification in 1961. I also captained the Canadian juvenile football championship team in 1964. What was interesting, in retrospect, was the fact that I did not consciously set any personal goals. I guess when you come from a working class family, your realistic life ambitions are to have a real job and to install telephones for Bell Canada, which I also did for two years. My father, until he died in 1991, told me that I should have stayed with Bell, not pursue a career as a gymnastics coach, because I might break an ankle!

One of the more humorous incidents concerning lifelong goals occurred when I was teaching a graduate sport psychology course in Brazil. I asked them, 'When you die, what do you think your friends and colleagues would have inscribed upon your gravestone regarding your goals and accomplishments'? The class was stunned by the question and when I asked them all to give me a response, the responses were hilarious. They never said: 'He worked harder than anyone on the team', or that 'He had his dreams and realized them', but were much more superficial. Obviously, these were people who had never thought of far-reaching, lifelong goals or exceptional goal-related accomplishments!

## 15.2 Self-confidence

Another element reported to be extremely important in the achievement of exceptional performance is belief, or self-confidence (Harris & Williams, 1993; Nideffer, 1988). As previously mentioned, self-confidence is another core skill that Orlick (2008) reported as being essential: 'The highest levels of personal excellence are guided by belief in one's potential, belief in one's goal, belief in the significance or meaningfulness of one's goal, and belief in one's capacity to reach that goal' (p. 112).

Self-confidence has been reported to be cyclical and variable. Individuals can believe in themselves more on some days than on others. Orlick (2008) revealed that individuals manifesting high levels of self-confidence often had a solid support network – that is, other people and loved ones who believed in them, and maintained a positive attitude towards their performances. These individuals received much positive and constructive feedback, drew out constructive lessons from training and competition, and regularly experienced improvements and successful performances.

It can be seen that commitment and belief in oneself are critical mental perspectives athletes need to develop and maintain their gymnastics performance to achieve their high level goals and skills. Moreover, it has been seen that goal-setting is an important skill athletes have used to enhance their self-confidence and commitment to excel. Other mental skills inherent to achieving success have also been identified in the literature, such as developing the *flow* state (Jackson & Csikszentmihalyi, 1999). First of all, flow is a state of consciousness where one becomes totally involved in the given activity, with the exclusion of all other negative thoughts and emotions. Flow has also been called *the ideal performing state* (Uneståhl, 1975), or *being in the zone*, and does not necessarily end upon winning.

The flow state occurs when gymnasts perceive a high level of challenge in a competition and in skill, and have performed and trained well to have the perceived competence and this is the situation where the state of flow *may* set in (Jackson & Csikszentmihalyi, 1999). This phenomenon does not always occur so often in gymnastics; however, it seems to increase with additional experiences. Gymnasts might feel that their performance is effortless, non-stressful and that time, distortions may occur so that the performance from a psychological perspective is either in a slow dream-like state or happens very quickly, or in *flow*. The occurrence of flow is affected by many factors, such as motivation, appropriate psychosomatic mental states, appropriate pre-competitive and competitive plans, and physical and mental readiness (Weinberg & Gould, 1999). The good news is that both skill levels and challenges are teachable and learnable mental skills, which can change with practice. But the source of flow is not always evident to the performer. But the ultimate aim for a gymnast and a coach is to appropriately balance both physical and mental training, which can eventually enable them to enter this flow state.

### *15.2.1 Implications of self-confidence for learning and performing in gymnastics*

A fundamental element in the mental skills foundation triangle is self-confidence, or belief. Often non-gymnasts, when asked 'What is an important mental concept central to success in gymnastics'?, report the term *self-confidence*. We can all remember in elementary school our colleagues who were picked out as being self-confident in math or reading and – usually, the older boys or girls – in sports. Self-confidence is developed by having achieved success in gymnastics, often before their peers could achieve a given skill. Self-confidence is rewarded by coaches, which, in turn, increases gymnasts' belief in themselves.

Prior to elementary school, I was fascinated on TV by the weekly hour of professional wrestling, which included midget wrestlers. I began practising and learned the single skill of a neck spring on our lawn, which was often done by the midgets to quickly kip up from their backs on the mat to their feet. It wasn't pretty and I must have looked like a four-year-old one-trick midget wrestler! But when I entered my Grade One physical education class, I repeatedly did them in front of the gym teacher on a wooden floor, much to his delight! I believe that this move was my first experience in developing my self-confidence in sport. But as is the case in any physical or academic activity, there are daily ebbs and flows of performance and it is for this reason we have peers or coaches to provide immediate feedback for less well performed movements and to reinforce what Orlick (2008) refers to as 'highlights' or small, daily positive events to improve self-confidence for the next day.

Setbacks and losses of confidence can also occur at the highest levels of performance with World and Olympic champions. When I was accompanying the Iranian national delegation to the Asian Games in Qatar in 2006, the tae-kwon-do coach directed me to Youssef, an athlete, for a discussion regarding his confidence This young man had won the silver medal in the Athens Olympics and was twice World champion, yet he had lost his confidence in the four bouts that were ahead of him in Doha. I met with him seven straight days and even on the day of the bouts. I was becoming exasperated and told him that he only had 24 minutes left to fight and had already accumulated 14 years of high level training, and achieved many Olympic and World victories.

After the first fight, which he easily won, I approached him in the warm-up gym with a sign on which I wrote: '18 minutes', after which I got up and left the hall. After each successful bout, I returned with new signs stating '12 minutes' and then 'Six minutes'. By this time, the whole Iranian team was laughing uproariously, including him. He won the gold medal, but he needed a bit of a self-confidence shove!

An example of this state of flow in gymnastics is illustrated by a former World and Olympic champion in gymnastics, whom I personally observed during the 1976 Montreal Olympics and to whom I later commented about my perceptions. In 2004, I had the privilege of participating in an FIG Coaching Academy with Nelli Kim, double Olympic gold medallist on the vault and the floor exercises in Montreal. During the Games, I was sitting in the best seats with my graduate students doing behavioural observations of the Olympic gymnasts (Salmela *et al.*, 1979). I recalled to her that she seemed in the psychological *zone* for exceptional performance, as described in the flow literature.

She told me the exact characteristics of her flow state: effortlessness, total contact with the routine, and distortions in time, as if competing in slow motion. I mentioned to her on the bus to the sport centre in Kuala Lumpur that I would be talking about this flow state that afternoon. She seemed surprised and replied to me: 'I always thought that this magical performance was because of God'. I kept this alternative hypothesis in mind as a possibility, but as a scientist, the flow hypothesis was more testable in scientific settings, rather than in religious musings.

In the preparation for the 1988 Seoul Olympics, during a three week training camp, the coaches and I came up with a confidence building exercise where all seven members of the team had to go through their routines twice successfully, without a major error by *anyone*, or else *everyone* had to repeat the whole process again. We also told them that we could delay supper if there were many misses. Since each gymnast's performance affected the number of repetitions for their team-mates, this exercise, both in self-confidence and commitment, permitted the whole team to go through 84 consecutive routines without a major error. On our rest day we played golf, and the gymnasts were still glowing. This stressful training exercise served us well and we achieved our best team finish ever in Seoul. Oudejans (2008) supported these methods and suggested that '...turning up the heat from the very first day of practice may be one of the most effective ways to immunize yourself against *blowing* it', or 'choking'. He continued: 'They're trained in how to play the game, but they don't train under pressure, so they will fail ' (p. 40).

Another non-gymnastic, but remarkable, example of self-confidence, was shown by a Canadian canoeist, Larry Kane. He had printed up in 1983 a business card on which was written, 'Larry Kane, 1984 Los Angeles Olympic Canoe Champion', which he distributed to everyone, and he won! Now that is an example of public self-confidence!

## 15.3 Commitment

Goal-setting may be 'one of the best performance enhancement techniques available in the behavioural sciences' (Burton, 1993, p. 469), but failure to develop an ongoing commitment to attain the set goals may diminish enhancement capabilities. Orlick (2008) perceived an athlete's level of commitment to be a critical ingredient for success. Niemi (2009) showed strong evidence that placebo effects in both medicine and sport can be powerful: they can positively affect commitment even though they are a sham, either medically or through psychological misinformation, like 'You are the best ever'.

In order to achieve excellence, individuals must possess or develop high levels of commitment, to the point where one's activity in a sport endeavour becomes the main life focus during a specific, and sometimes relatively short, period of time in sport (Orlick, 2008). However, being totally committed may also have detrimental effects on performance, leading to exhaustion or athletic burnout (Weinberg & Gould, 1999). To prevent overtraining, Orlick suggested balancing commitment with appropriate recovery periods, and joyful activities, not related to work.

Researchers have shown that there are several ways to increase the levels of commitment in a chosen activity. Having an inherent passion, or love for a given sport, leads to higher levels of commitment. Perceiving goals to be worthy and achievable, and believing in oneself, also have been shown to be characteristics of highly committed individuals (Orlick, 2008).

Orlick discussed the important roles it plays in enhancing athletes' self-confidence and commitment levels. Furthermore, according to Harris and Williams (1993), the levels of commitment of athletes can be heightened when certain sacrifices are made, when the time and effort invested is acknowledged and supported by their family and team-mates, and also when their commitment is publicly displayed.

In the attempt to explain the attainment of expert performance, Ericsson *et al.*(1993) proposed a model which postulated that 'The primary mechanism creating expert-level performance in a domain is deliberate practice'. Deliberate practice was defined as an effortful activity that is motivated by the goal of improving performance. Of interest is that, unlike play, deliberate practice was perceived as not being inherently motivating, and unlike work, it does not lead to immediate social or monetary rewards. Furthermore, the length of time required to reach an expert level in a particular domain was estimated to be at least 10,000 hours of meaningful, goal-directed practice, which virtually defines commitment.

The three commitment questions that best discriminated between the international and the national group on the OMSAT-3 were: 'I am willing to sacrifice most other things to excel in my sport'; 'I am committed to becoming an outstanding competitor'; and 'I feel more committed to improve in my sport than to anything else in my life' (Durand-Bush, *et al.*, 2001). This is not surprising since, as has been previously mentioned, researchers have found that elite athletes are extremely dedicated individuals who are willing to do almost anything to become the best, even if this means sacrificing everything else that is important to them for a certain period of time – and unfortunately, as we have recently observed, has led to cheating with the use of illegal substances (Orlick, 2008; Orlick & Partington, 1988).

### *15.3.1 Implications of commitment for learning and performing in gymnastics*

Commitment to practice and competition is not just the task of the gymnast: consideration must be given to the parents and the siblings of aspiring gymnasts and to the coach. In large urban centres, where most gymnastic centres are centralized, because of the high costs of equipment and coaching, young gymnasts are most often transported to and from the gyms by their parents, and if their parents are working, by caretakers or grandparents. If the child has siblings, it is rare that they will also participate in the same niche or activity, since their time resources are already stretched thin, which makes the social dynamics and the transportation more complex (Côté 1999). Thus the level of commitment is not just the decision of the aspiring gymnast, but also of the siblings and the parents.

At the Gymnix Club in Montreal, one of the city's most successful women's gymnastics clubs, with multiple Olympians and world performers, a unique procedure for gymnasts' selection was undertaken by the founding president and head coach, Nicole McDuff, during the 1970s. She would invite aspiring young girls and their parents to a three- to four-hour training session with her elite team, and after the training session, she would independently ask both the parents and the young candidates whether they wished to endure the necessary parental transportation to and from the gym, six days a week, for three to four hours of training, as well as the weekend competitions over the next 10 years! In more than 50% of the cases, this question regarding commitment was negative and this solved much wasted time which would have been spent with someone who would have eventually dropped out of the sport, whether it be the gymnast or the parent. To support this point of view, the best-discriminating OMSAT-3 question of all was 'My sport is the most important thing in my life'. As an adolescent or a young child, this may be true, but it seems obvious that this perspective would change with the nature of the professional and daily activities of the family.

But commitment extends at times beyond personal commitment to that of a gymnast's team or country. I was present at two events at the 1976 Montreal and the 1988 Seoul Olympics where this commitment level went way beyond what could have been imagined. In Montreal, the Japanese team had already created a dynasty in the men's team, with team gymnastics titles in the 1960s and 1970s. By the time of the Montreal Games, Japan had won four consecutive team gold medals. During the team finals, Shun Fujimoto broke his knee cap on floor exercises. Rather than withdrawing from the meet, he masked his injury on his final events, the pommel horse and rings. On the rings, Fujimoto scored a 9.7, after landing his full-twisting double back dismount with his broken knee cap, on the other foot! His score helped the Japanese earn their fifth consecutive team gold, and he is still revered in Japan for his selfless commitment to the team. Another Japanese gymnast who demonstrated this same commitment was Sawao Kato at the 1974 World Championships in Varna. He was Olympic all-around champion in 1968 and 1972, but during his horizontal bar routine, he dislocated his shoulder and his arm was out of its socket. Despite this, and with the encouragement of the fanatical Japanese fans, he attempted unsuccessfully to remount with one arm dangling from his shoulder. He tried, but logically he could not continue, and failed, not in his commitment or courage, but for obvious medical reasons.

Another example of what could be arguably called team commitment, mental toughness or stupidity, occurred with Bobby Baun in Canada during a 3–3 tie in the sixth game of the 1964 Stanley Cup ice hockey final. The Toronto Maple Leaf's defenceman was hit with a hard shot which broke his ankle. He returned to the trainer's clinic and they told him that it was surely broken. He said that it could not hurt more, so he told them to just inject him with pain killers, and wrap it tightly, so that he could return for the overtime. While not being known as a great goal scorer, he scored the winning goal, and the pointed out that 'This was the greatest *break* in my life'! Now you be the judge in which of the above categories he should be placed: either committed, tough minded, or stupid.

The final instance was a bit closer to home for me, since I had been involved with the Canadian team and in mental skill consultations for more than 10 years. I had met Philippe Chartrand 10 years previously when we were conducting the T–T, or talent identification programme for men gymnasts across Canada, using more fixed and/or genetic variables. But honestly, anyone could have 'eye-balled' out this strong, flexible and engaging gymnast as a 'talent' in gymnastics without any other scientific information. Philippe progressed across his career and became the world champion on the horizontal bar, in the 1983 Edmonton FISU Games at the age of 21. During the six week preparation for the 1988 Seoul Olympics, we went through a number of competition contingency plans, one of which was what the team should do if one of their team-mates was injured during the competition. The consensus was to step over the body and start focusing on the next event.

During the team finals at Seoul, Philippe dislocated his knee on vault and everyone thought that he was through. But the team did as trained, ignored him, and went on to warm-up for the parallel bars. But Philippe was no ordinary gymnast, he was the team captain, made of steel and when the other gymnasts were warming up, he shook his leg a bit on the sidelines, put on his white pants and then proceeded to the parallel bars, which stunned the whole team. Since he was the last and best man up for the Canadian team, he went through his full warm-up and finished with double and triple somersault dismounts on the parallel and horizontal bars, landing on one leg! Because of him, the team's personal worry diminished and Canada finished in 9th rank, Canada's highest placing ever! Now this was commitment to both the team and the country.

## 15.4 Implications of the mental skills foundation triangle for learning and performing in gymnastics

A number of important lessons can be drawn by considering the foundation skills and their crucial interactions. First of all, unless the foundation skills are assessed as being relatively high, there is little chance of achieving success in gymnastics or any other realm of achievement. Concerning goal-setting, if you do not have a map before you, either drawn up by you, your coach or both of you together, chances are that you will end up somewhere else. Start out with what would be your ultimate dream, such as making the national squad, representing your country at the Worlds or Olympics or winning a medal.

The second element of self-confidence or belief is the glue that brings together the three fundamental components. Recent evidence on national and international athletic medallists, including gymnasts, has nicely tied together these three concepts when considered across their levels of expertise in sport. I have met some lower skilled gymnasts whose dream was to make it to the Olympics, and when they got there either did not compete or faked an injury. Obviously, their goal-setting was inappropriately low: of just making team and going to the opening ceremonies, but their pride in their accomplishments was not very high. If they were committed to that dream, and their goal was to make it to the top half of the draw or to one final, this dream goal would have been acceptable. It is clear that for exceptional performance in any domain to be achieved, the gymnasts must be committed to extensive practice over many years (Ericsson, 2007). Currently, the commitment levels to practising are up to 30–36 hours a week of conditioning and training, with only short summer holidays.

## 15.5 Psychosomatic skills

### *15.5.1 Stress control*

Stress is an intrinsic component of training and competition. Research has shown that negative reactions to stress, or competitive pressure, can be detrimental to gymnastic performance, and that conversely, positive reactions to stress, arousal or nervousness can lead to enhanced results (Rotella & Lerner, 1993). Murray (1989) conducted a study in which athletes were asked about how they interpreted their pre-competitive arousal levels. Over 70 per cent of them reported that they enjoyed the nervousness associated with competition, that it helped their performance and was a good indicator of their readiness to perform. Rotella and Lerner have thus stressed the importance of developing effective ways to respond to stressful situations which could potentially limit the achievement of a gymnast's goals.

Much research has been conducted on the topics of stress, anxiety and arousal (Gould & Krane, 1993). The three terms have often been used interchangeably in the sport psychology literature. Martens (1977) defined stress as

> A process that involves the perception of a substantial imbalance between environmental demand and response capability, under conditions where failure to meet the demand is perceived as having important consequences and is responded to with increased levels of the A-state [state anxiety].
>
> (p. 9)

Spielberger (1966) made an important distinction between state and trait anxiety. State anxiety was defined as a situation-specific emotional state that reflects the perceived feelings of apprehension and tension, which are associated with increased or decreased arousal. Conversely, trait anxiety was defined as a stable behavioural predisposition to many situations, which are perceived as threatening. Various theories have been postulated to explain the relationships between arousal and performance, or anxiety and performance (Gould & Krane, 1993). Recent findings have led researchers to believe that the multidimensional theory of anxiety contributes greatly to the understanding of the anxiety–performance relationships. This theory predicted that cognitive and somatic anxiety affect sport performance in different manners (Burton, 1988). More specifically, it suggested that there are strong negative linear relationships between cognitive state anxiety and performance, and a less powerful inverted-U relationship between somatic anxiety and performance. Gould and Krane raised a need to conduct more research in this area before valid inferences can be made from the assumptions of this multidimensional theory of anxiety.

Jones and Swain (1995) reported some desirable dimensions of high stress levels in sport if the athlete considers these emotional states to be facilitative, rather than debilitative, for successful performance. Thus, in exactly the same context, successful athletes can reframe a potentially stressful threat into an exciting challenge. In a qualitative study with high level athletes, Jones, Hanton and Connaughton (2002) considered that mental toughness allowed athletes to '... cope better with your opponents with the many demands,...that sport places on a performer, and be more consistent and better than your opponents in remaining determined, focused, confident, and in control under pressure' (p. 213). Anshel and Payne (2006) termed this state 'managed intensity', specifically referring to effective strategies for athletes in the martial arts. They coined the expression 'challenge appraisals', where the fighter is externally focusing upon heightened vigilance in the martial arts situation, rather than internally worrying about possible performance outcomes, such as winning, losing or even just qualifying for the next round of competition.

Rotella and Lerner (1993) made some insightful comments when considering responses to competitive pressure:

> The more similar the abilities of the competitors and the greater the importance of the event, the more likely that high levels of pressure will be experienced by athletes facing that challenge. They must be able to consistently perform at or near peak levels when exposed to the highest levels of competition.
>
> (p. 528)

Murphy and Jowdy (1993) found that imagery and mental practice techniques are important components of stress management. Lazarus and Folkman (1984) found that subjects used imagery interventions to become more familiar with effective strategies for coping with stress. In other studies, imagery techniques were successful in reducing different types of anxiety, such as medical anxiety and test anxiety, and also in changing sports behaviour.

Recently, Svoboda (2009) outlined research on an extreme lack of stress control, or *choking*, simply defined by Baumeister (1984) as performance decrements under pressure circumstances. Beilock and Gonso (2008) instructed groups of both novice and expert golfers either to take their time and putt slowly or to putt quickly. When asked to putt quickly, the novice golfers performed poorly, while the expert golfers putted better. They speculated that an over-learned skill of an expert may suffer from too much conscious thought, and when they had to pay attention, rather than just perform, they started worrying and monitoring their performance, and in effect began choking.

Hardy (1990), who has worked with the British national gymnastics team, developed a model which explained the complex effects of *physiological arousal* (muscular tension, sweating, increased heart and respiration rates), which has been shown to be beneficial for increases from a sleepy to an optimum emotional state of performance. But with further increases in arousal, it causes performance decreases. This has been termed the inverted-U hypothesis, which has been well known for more than 100 years (Yerkes & Dodson, 1908). The inverted–U phenomenon can be demonstrated in anyone's daily life activities. For example, when you wake up in the morning, your physiological state of arousal is at its lowest, and this would not be a good time to do your income taxes, since your performance would necessarily be low. However, once you get up and about, your physiological state would improve and you start thinking more clearly. If, however, you decide to go for a vigorous run, and then, upon returning, immediately do your taxes, this might also not be the optimal period, because of your fatigue. This physiological activity level would follow a bell-shaped curve, and if you had no further stresses, it would slowly return to the early morning state and you would fall asleep.

However, *cognitive anxiety* is another issue, since it deals with worry, doubt, fear and other external pressures. In gymnastics, it can be related to worrying about physical and mental readiness, the importance and consequences of doing well, making the team, the strength of the competitors or even the nationality of the judges. However, when *physiological arousal* is combined with *cognitive anxiety*, or states of doubt or worry, a threshold is reached and the arousal curve drops dramatically, unlike the smooth physiological arousal bell curve – it's like a pool ball falling from the table. There is a gradual increase in the curve and then a 90-degree fall. This results in immediate performance decreases. Hardy (1990) therefore introduced the idea of the *catastrophe model* (which could provide another explanation for choking in stressful situations in sport). This state can be reduced after physical relaxation and the mental restructuring of the gymnast's thoughts, in which confidence and control are regained.

### *15.5.2 Implications of stress control for learning and performing in gymnastics*

An example of the application of the catastrophe model could be seen if a gymnast was asked to do a kip to a perfect handstand on the bars, swing down and do a free hip circle back to the handstand position. In normal conditions, the gymnast would be in a calm state of *physiological anxiety*, but if asked to do a series of five or ten, each ending in a perfect handstand position, a well-trained gymnast could probably perform this with ease. If, however, the gymnasts were instructed to do 10 free hip circles to a handstand perfectly, with failure to do so meaning their elimination from the Olympic team, *cognitive anxiety* would rise considerably, and their performance might break down in a catastrophic manner, because of the worry caused by this increased cognitive anxiety.

In 1994, I was consulting with the Canadian men's and women's biathlon team in preparation at a training camp for the Lillehammer Olympic Winter Games in Norway. On the women's side, Canada had Myriam Bédard, who was the overall season's champion for both the 10 and 20 km races, which included cross-country skiing and target rifle shooting every two kilometres. She asked me in what sport I was principally involved and I said gymnastics. She then made a very insightful comment: 'So your gymnasts deal with real fear stressors and biathletes with real pain', two very different sources of stress. What was of interest to me was that of all the members of the team, she was the only one to take notes during the interventions!

It is very difficult at this point to ascertain the nature of the stress reaction, as being a true mental skill. In one instance, it was found to discriminate well between athletes (Salmela *et al.*, 2009) and in another (Durand-Bush *et al.*, 2001) it did not. In both cases, it was found to have weak correlations with the other scales in the inventory, except for refocusing; however, the scale still yielded acceptable levels of internal consistency. This scale has not been found in any of the other multidimensional inventories, but this does not mean that it is not an important skill or capacity that athletes need to develop to achieve high performance levels. Examples of two items for the stress reaction scale were: 'My body becomes unnecessarily tense in competitions' or 'I have experienced problems in my performance because I was very nervous'.

Another explanation for the inconclusive results could be that the scale is sport-specific, that is, it may apply more to athletes involved in sports like gymnastics, alpine skiing, luge, bobsleigh and whitewater kayaking, in which physical stress is a major factor in training and competition. Yet, items in this scale such as 'I am afraid to lose' and 'I am afraid to make mistakes' apply to every athlete, not only to those participating in stress inducing sports.

In 1964, I was on a three day train ride from Montreal to Winnipeg for the Canadian national football championships, where I was one of the team captains. Upon arrival the team was told that we would have an evening practice in the dark, snowy conditions on the night before the game. We were exhausted, stressed and not in a good humour, but just as our bus was about to arrive at the field, one of our assistant coaches said 'Lads, this will be a fun practice before our war tomorrow'. He then took out his upper set of dentures, threw them into the air and then, after four to five rotations, caught them in his mouth, and the whole delegation cracked up.

We had a short practice in freezing conditions, but perhaps this humorous moment helped us win the national championships, with the stress conditions reduced by this coach's antics.

What must be remembered is that gymnastics is a sport that is unnatural for human beings. The human vestibular, or balance, system in the inner ear is designed for individuals to remain upright, balanced in a standing position with the head on top in the normal anatomical position. However, many gymnastic movements, with the handstand being the most obvious one, require that the gymnast perform upside-down, which, as a beginner, is stress inducing. For this reason, many of the early photographs of Olympic champions from the 1920s until the 1950s showed the gymnasts in an arched position in the handstand position, so that the head could remain in a somewhat normal, but less stressful, vertical position. Therefore, gymnastics cannot be considered, in relation to stress, as a normal human sport activity, such as running, walking, throwing or jumping. It is inherently biologically, anatomically, and psychologically stressful!

### *15.5.3 Fear control*

On this subject, psychologist David Feigley (1987) reported his far reaching observations on fear in gymnastics:

> For some gymnasts, fear is the major psychological barrier preventing learning and success. It prevents the learning of new skills and retards the improvement of already learned ones. It causes athletes to feel helpless and out of control of their lives. Fear erodes athletes' feelings of self-worth and competency and destroys the fun of the sport. Fear even ended the careers of otherwise successful gymnasts who quit because they were unable to cope with its constant presence.
>
> (p. 13)

However, overcoming fear is one of the major reasons for certain types of individuals to be attracted to the sport of gymnastics. Feigley outlines that gymnasts must '...perceive that they can effectively assume individual responsibility and personal control over the fears they encounter in their sport. Successful gymnasts, feel in control of their lives' (p 13). However, fear cannot be eliminated from high risk sports such as skiing, ski jumping, motor sports and gymnastics. It is fear that attracts the moth to the flame – close enough is fun, but too close can be a disaster. Many gymnasts try to eliminate fear, but it is a natural reaction to the risky skills that gymnasts perform and is actually a necessary component for the advancement of gymnastics.

Cogan (2006) related: 'Fear helps a gymnast maintain enough focus to perform difficult skills safely. Therefore, a gymnast's goal should be to work with fear, rather than to eliminate it' (p. 644). Often fear is the result of carrying too much excess mental baggage, or negative thinking, from daily life, by thinking too much. Gymnasts should learn to drop this luggage mentally, by visualizing opening their hands and lowering the luggage to the ground. Another strategy, suggested by an Australian, Peter Terry, is to maintain their favourite attention grabbing music on their IPods until just before the competition, a strategy that eight-time Olympic gold medallist Michael Phelps followed during the 2008 Beijing Olympic Games in swimming.

Feigley (1987) astutely points out:

> A gymnast who is never afraid is either foolish or ignorant. Sophisticated gymnasts experience fear on a regular basis. Those who are successful at their sport have developed

> the means of dealing with their fears so that the risks are controlled. Unfortunately, some athletes try to deny their fears because they: a) view fear as a weakness; b) are afraid to look foolish; c) or believe that they are the only one in the group who is afraid.
>
> (p. 14)

Herein lies the role of a skilled coach or mental trainer.

### *15.5.4 Implications of fear control for learning and performing in gymnastics*

When I was an undergraduate student, I was extremely fortunate to have participated in a week-long course in life guarding and swimming from the Emeritus Professor Murray Smith from the University of Alberta. In 1970, he was the first person who took a humanistic view of the relationships between fear, excitement and learning in sport. He related how a five-year-old boy was terrified about jumping into the swimming pool from a one metre diving board, after having done so successfully from the deck, three centimetres above the water's edge. The young lad was frozen on the board, and Murray said ; 'Max, it's the same thing, but its only a little bit more exciting, and just like from the pool's edge, and you will float back up to the surface'. Max said, 'Yes it's very exciting, so I will try my best', and he jumped into the pool! Murray turned Max's fear into excitement, which had the same level of activation, but this resulted in a mental transformation in Max's mind, from fear to excitement. It was a great success for both Murray and little Max.

This is what Kerr (1997) referred to as *reversal theory*, according to which it is not only the degree of fear or stress control that may affect performance in gymnastics: there are individual differences which individuals may sustain in their *interpretations* of fear or stress which may play even larger roles. We know that gymnasts may be highly activated, as was the case for Max, and initially this state could be interpreted as being negative and stress inducing. However, Murray reversed the perception of Max's mindset from being fearful to one of pleasant excitement, and often a good coach or mental trainer can initiate these changes.

I knew an American gymnast who, as a high school student, was in the top three in the world in competitive trampoline. During the summer time, he was a small plane crop duster in the mid-west United States and I am sure that his ultimate demise was due to the fact that he was trying to accomplish the then unaccomplished triple twisting triple back, in his small aircraft, before attempting it on the trampoline, but he crashed. His high skill level and low level of fear resulted in his death in a midwest corn field. This points out that the lack of fear may not be desirable, even though the rush stimulated him to surpass the normal standards in gymnastics, through its pure excitement.

I shared a similar, but less dramatic experience in gymnastics when in 1989 I accompanied a group of 13–16-year-old male gymnasts in Lilleshall, England, with the coaches' tasks of submitting them to intense technical training, and I provided mental training for six coaches and 12 athletes who were our hopefuls for the 1988 Seoul Olympics. Besides the mental training sessions, the mission was to learn and develop the new prescribed skill difficulties to help qualify Canada for the Olympic Games.

Learning new elements necessarily meant dealing with new fears about being injured. When I did the theoretical part of dealing with the various types of fear, e.g., realistic versus unknown fear, the class became quiet. We specifically were dealing with the Tkatchev, a relatively new skill on the horizontal bar. The movement was quite scary, since it was initiated by an overhand giant swing and about three quarters from the initial vertical starting position, the gymnasts' body sharply arched and then flexed, which resulted in a counter-rotation of the giant swing and propelled them backwards over the horizontal bar with no visual guidance and, hopefully, a successful re-grasp on the other side of the bar. The frightening elements were contacting the bar with their body. When I specifically asked about what their greatest fear was regarding learning the Tkatchev, one bright gymnast said the following: 'My teeth on the steel bar!'

Later that day, in this fully secure, modern training facility, our young hopefuls did the lead-up exercises, which included the pre-release 'timers' of a beat and an arch, but no one let go of the bar! Then one lanky French-Canadian, Benôit, after doing 25 timers, went for it! He released a bit early and went about two metres above the bar, dropped and hit the back of his thighs on the bar, which flipped him harmlessly backwards into the foam landing pit, and he surfaced laughing and grinning.

The worst fear-inducing alternative had occurred, and he was still alive, happy and uninjured! Benôit had turned his fear into team excitement, and transformed these awfully negative emotions into a higher positive state of excitement within 10 seconds. Within the 20 minutes that remained before our lunch break, this transformation spread among all of Ben's team-mates, and every one of our young squad either released their grasp over the bar and landed in the foam pit, or successfully re-grasped the bar for another attempt. Such was the beautiful mental link between fear and excitement.

However, there are also a number of non-rational fears involved in gymnastics which can be best resolved by the coach discussing the use of either the biomechanics of the movement or its possible performance consequences:

> Another type of mental block is losing a skill that a gymnast has performed alone or in competition without difficulty. One day she feels a little disoriented and does not throw the double full on floor quite right, or her timing is off...From there her performance deteriorates until she cannot do the skill at all.
>
> (Cogan, 2006, p. 645)

So how does a skilled gymnast transform a well learned skill after years of practice into a fear-inducing element? To help illustrate one example of the above situation, I was invited to consult with a women's coach whom I knew, whose daughter was the provincial junior champion and a champion in the nationals in one month. She told me that her daughter was afraid of doing a back walkover on the beam, a skill that she had been doing in competition for eight years. When visiting the gym, I asked the gymnast to do a back walkover 10 times on a mat. She did them perfectly. I then asked her to do the same on the low beam which was about 25 cm from the floor, which she also did wonderfully.

Then I heard a scream from across the gym from her mother, who was also her coach, telling her that she had a competition in one month, that she had done this movement thousands of time on the high beam, and that she needed this movement in the compulsory exercises. I stopped my intervention and approached the mother/coach and said: 'Do you want to lose a gymnast or your daughter? She must be coached by someone else'!

Most gymnasts are proud when they accomplish a new element and then excitedly return home to share their success with their mother. But when the gymnast's coach is her mother, there is shame, fear, and no place to hide. Ten years later, they both approached me at a national championships and thanked me for my very brief, but helpful intervention. She was from then on coached by another colleague and received a full gymnastics scholarship to an American university, where she did very well, and she loved the experience. Such a mental skill learning effect for such a small, but important intervention!

### 15.5.5 Relaxation

Relaxation is a method often employed to decrease arousal or worry. Although several different types of relaxation techniques exist, they are all variations of those developed by Jacobson (1938). Zaichkowsky and Takenaka (1993) made the following statement regarding Jacobson's progressive muscular relaxation PMR) technique:

> It is clear that the mastery of the Jacobsonian relaxation techniques results in reduced levels of anxiety, muscular tension, and physiological arousal. Research on the efficacy of the specific PMR technique for improving athlete performance, however, is still quite sparse.
> (p. 521)

Effects of relaxation have often been studied in combination with other arousal control techniques, such as deep breathing, meditation, and cognitive techniques including imagery and self-talk (Zaichkowsky & Takenaka, 1993; Uneståhl, 1975). Relaxation techniques have not only been used to regulate arousal, they have also been used to control anger, reduce muscular tension and promote assertiveness, concentration and confidence. Relaxation techniques can be divided into two different categories: *muscle-to-mind* and *mind-to-muscle* techniques (Harris & Williams, 1993). Jacobson's PMR, which involves tensing all muscle groups progressively before relaxing them, falls into the former category. However, this method is lengthy and somewhat boring for gymnasts, and does not include essential cognitive rehearsal. On the other hand, transcendental meditation would be categorized as a mind-to-muscle technique.

According to Harris and Williams (1993), relaxation skills must be practised on a regular basis. Although some individuals may take longer than others to develop these skills, most people are able to observe improvements after a couple of sessions of practice. Harris and Williams emphasized the importance of being able to relax completely and quickly. They reported that through deep relaxation, athletes can detach themselves from the environment, allow their central nervous system to regenerate physical, mental, and emotional energy, and create a base for learning *quick* relaxation. They defined quick relaxation as the ability to relax within a short period of time. It, also, can be an effective strategy to regain a full focus during competition, and to return to a balanced, controlled state of mind after competition.

### 15.5.6 Implications of relaxation for learning and performing in gymnastics

A number of psychologists have used the Jacobsonian PMR technique with gymnasts, but since they are required to relax all muscle parts through contractions and relaxation, without cognitive components, and due to its length, this requires something that serious athletes have little of, and that is spare time.

A tried and tested combined relaxation script has been successfully applied with thousands of athletes in a number of countries and it take less than 20 minutes. It includes a brief muscular contraction, followed initially with a script that is spoken by either a coach or a mental trainer (Uneståhl, 1975). Initially, with this muscle-to-mind procedure, the gymnasts should preferably be lying down barefoot on a comfortable surface, such as the floor exercise mat, with comfortable, loose clothes, with their feet apart and their eyes closed. This form of cognitive–behavioural relaxation has proven to be effective with single subjects or with large groups.

It initially involves a single, but intense, muscular contraction of the non-dominant hand and the taking in of a long, deep breath which they must hold as instructed by the mental trainer in a loud, forceful voice while calling out

*Harder, harder, hold it, hold it*

for 10–15 seconds; then the command is given to relax, spoken in a lower and calmer tone.

The gymnasts are then introduced to the mind-to-muscle cognitive or mental component, and are instructed to:

*Notice the differences between the contracted and the uncontracted arm.*

This directed suggestion element begins in a quiet voice. The script continues:

*Notice how the contracted arm feels warmer and heavier than the other. It presses against the floor and is warm and heavy, warm and heavy.*

Now the first imagery part is initiated.

*With this warm and heavy sensation, imagine it as a bright, red, warm ball of light. Slowly, make it move up your arm to your shoulders which are now pressing against the floor and which are also warm and heavy. Now move the ball of light up to your forehead and the wrinkles in your brow will disappear and your head now presses against the ground and is warm and heavy. The ball now moves down to your mouth which opens slightly and the down the other arm, which is now also warm and heavy. Enjoy this calming period where your upper body is pressing against the mat and your body is warm and heavy. Now have the ball of light move back up your arm, to your back which presses against the mat and is warm and heavy. You are now feeling more and more comfortable, certain and in control.*

This is the positive suggestion phase, which can be repeated every one to two minutes.

Using these same instructions, in an even calmer tone of voice, you say:

*Now move the ball of light and the accompanying feelings of relaxation down through your buttocks, thighs, calves and feet,*

using the same quiet and soothing instructions. In my experience, this whole procedure, with practice, takes only seven to nine minutes. If they have just eaten or worked out, they often report that my instructions are sometimes heard and then wane in and out. I will then direct their attention to their breathing, another effective relaxation technique, and say

*Notice that when you inhale there is now an increased tension in your chest and when you exhale you are more relaxed, relaxed and heavy.*

If they are extremely tired after an extended training camp, I will stop here and let them recover into a deep sleep.

The mental skill of imagery is introduced below, in the cognitive skills section, but for continuity's sake, now is the time to include some other imagery skills. Another script is next introduced:

*Imagine that you are in front of a set of 10 stairs. As you descend the first stair you become even more and more relaxed. Now at the second one, you get deeper and deeper and you become more and more relaxed, in control, certain and comfortable.*

Continue this until they have reached the tenth stair.

Now the phase of the gymnastic-specific imagery begins. This suggestion is then given:

*Look in front of you and you will see a large, sky blue bubble. This is your personal mental bubble which keeps stress, worries, and other distractors (e.g., other gymnasts, coaches or parents), away from you and you can use it either in training or competitions. It is nicely decorated and in the middle is a soft comfortable reclining chair where you can now sit, or lie down, and practise some elements in your routines. In your bubble, these movements are always carried out effortlessly and perfectly, without fear or anxiety.*

Let them enjoy this experience, which they do, and then a number of gymnastic skills can be introduced which should help them imagine with confidence and certainty.

For the women, the sequences may include approaching the beam in a competition, mounting and performing with ease, on a beam which appears wider; performing their first difficult elements, or the final tumbling sequence with a stuck dismount, which they can easily repeat 10–20 times within a short period of time. For the men, it is usually the pommel horse where they start, and the instructions include executing flat double-leg circles, with fast hands, finally ending with a high, controlled and stuck dismount.

I will then bring them back up the stairs with instructions to become increasingly awake and aware as they go up each successive stair. I will say

*With each stair, you will become more and more awake and feel more comfortable, sure and confident.*

The number of each stair is given and they are told that they are becoming increasingly awake and aware. From the eighth to the tenth stair, I ask them to move their fingers a bit, which is sometimes quite difficult because of their deep relaxation, then slowly move their ankles and bend their knees. They are then asked just to lie there and enjoy this altered state of calm control, and when they are ready, to sit up slowly. The whole process takes from 14 to 17 minutes and the gymnasts really enjoy it, which I always discovered in the short debriefing and feedback sessions held immediately afterwards.

After two to three sessions, I attempt to remove myself from the interventions because I do not want them to be dependent upon me, and I ask them just for the contraction, then have them imagine the red ball of light to relax their body, then to concentrate upon their breathing, go down the stairs and enter the bubble to rehearse. When they are ready, they should begin to reactivate themselves. I have done this whole process with over 200 athletes from 15 sports in Iran in a number of uncomfortable environments, such as sitting on hard chairs in a classroom, and there was only one athlete who could not, or would not, release, let go, and relax.

More than a decade ago at a national Canadian age group training camp, I carried out these exercises with about a dozen 12–15-year-old boys and after the relaxation session, asked them to imagine themselves, while within their bubble, approaching the devilish pommel horse, with full confidence, and to mount and to do 20 full, flat, effortless double-leg circles and then to do a simple dismount. On this first day of mental training, without instructions of what to do, except to follow the script, every gymnast, once in their bubble, began wiggling and circling on the floor until I gave them the verbal cue to dismount! I have had similar experiences with girls walking along an imaginary wider beam and doing split leaps.

It is also possible in critical situations to use the *one second relaxation technique*, which is often seen in close-ups of players in ice hockey or soccer penalty shoot-outs: just before performing, the shooter or the goalkeeper takes a single deep breath and expires fully. While not as dramatic as the above techniques, it does physically relax the shoulders and helps adopt a positive focus, rather than one based upon worry. This is also clearly seen in gymnastics before mounting on an apparatus, or before a final tumbling run on the floor exercises.

When I was with the men's national team at the 1995 World Championships in Sabae, Japan, one of the gymnasts, as we were walking by an old temple, asked me if the team could carry out this relaxation and mental practice that they had learned 10 years before, and which they had practised over the years. They still remembered the positive sensations of these mental training protocols, while being within this peaceful and spiritual environment.

### *15.5.7 Activation*

Sometimes athletes are under-aroused, or are mentally flat, before or during a competition. In these instances, energizing techniques would be most effective to increase their chances of achieving successful results in gymnastics. Many energizing techniques have been used by coaches and athletes (Anshel, 1990; Harris & Williams, 1993), despite the limited research on their efficacy (Weinberg & Gould, 1999).

Energizing techniques used in the past have included the use of rapid breathing techniques, running, stretching, yelling out loud, or listening to stimulating music or videos, or energizing imagery, verbal cues, pep talks, and doing energetic pre-competitive workouts (Zaichkowsky & Takenaka, 1993). Athletes have also been known to *psych themselves up* by drawing energy from their environment – that is, from the crowd, their opponents, team-mates, flag or national anthem. Zaichkowsky and Takenaka have suggested that athletes may also energize themselves by transferring negative emotions such as anger, fear, disgust and contempt into positive emotions such as excitement about their achievable challenges, which injects positive energy for their performance goals.

It is believed that certain factors have to be considered before arousal-regulating techniques are implemented and effective. According to Zaichkowsky and Takenaka, it is important for both coaches and athletes to develop a sense of awareness that will allow them to detect if and when arousal levels need to be altered. Secondly, coaches and gymnasts need to be aware of techniques which are most effective for each gymnast. Coaches have to realize that there exist individual differences in gymnasts' responses to arousal-regulation techniques.

### *15.5.8 Implications of activation for learning and performing in gymnastics*

It has already been shown in Figure 12.1 that the movement dynamics between the various events in gymnastics sometimes require a calm mental state, which should occur on the balance beam or the pommel horse, and at other times effort, as in the vault and floor exercises. However, it was found with the OMSAT-3 assessments that most athletes had no problems activating in competitive situations, but needed to learn to relax, both physically and mentally.

For example, the sport of weightlifting provides some most interesting examples, since the athletes need to exert maximum power within a couple of seconds. The coaches often scream aloud at their athletes, smack their faces and give them smelling salts to pep them up to lift the maximum weights. In gymnastics, these procedures may be effective on events which require power, but not so on those which require control of the finer movements for optimal performance.

I have seen a video of Mike Tyson and his famous coach, Gus D'Amado, prior to an amateur fight in 'Iron Mike's' boxing career, and he was crying in front of his coach. He was a boxer who would be feared by all of his rivals in professional boxing, but he could not get ready and activated for an amateur fight! I have also already described the lack of mental energy, or activation, of a two-time Iranian World Champion in tae-kwon-do and a silver Olympic medallist prior to an international competition. So, activation levels cannot be taken for granted, even at the highest levels of sport!

### *15.5.9 Implications of psychosomatic skills for learning and performing in gymnastics*

Emotional or psychosomatic skills are often the most dominating aspects of success in gymnastics. From my own perspective, fear was the greatest factor which limited my performance. During the 1960s and 70s, there were few gyms which had eight inch landing mats, hand grips for the horizontal bat were primitive and were not constructed to maintain contact with the bar, and there were no three to four foot landing pits – just a few two to three inch mats, which made attempting difficult tricks very dangerous. I have flown off the horizontal bar a number of times and struck my head upon a wooden floor. This poor spotting equipment did not motivate me to try risky exercises so often.

Doing tumbling on wooden or tile floors always limited doing somersaulting movements, mostly to competitive situations, and even the one inch foam pads were not that much better. Thus, in my case at least, fear was a limiting factor. I had no problem in relaxing and energizing, but fear and stress reactions, definitely limited my performance, and I applaud those gymnasts at the international level who went through these events with success.

## 15.6 Cognitive skills

### *15.6.1 Imagery and mental practice*

The terms 'imagery' and 'mental practice' have been used interchangeably in the sport psychology literature and thus will be discussed together in this section, until they are either experimentally or functionally shown to be independent variables. Both concepts are normally practised outside of the competitive context, while the mental skills of focusing and refocusing are typically executed during training and/or competitions.

Murphy and Jowdy (1993) emphasized the importance of carefully distinguishing between the two terms. Corbin (1972) defined mental practice as the 'repetition of a task, without observable movement, with the specific intent of learning' (p. 94). On the other hand, Suinn (1993) associated mental practice with techniques which included thinking about a skill by visualizing or *feeling* it. This involved self-talk throughout the steps of a skill, imagining oneself or another individual executing a movement, and incorporating auditory, proprioceptive and emotional elements, while visualizing the perfect way of performing. With this definition, it was specified that mentally practising does not imply engagement in imagery or mental rehearsal. In the research and interventions using the OMSAT-3 (Durand-Bush *et al.*, 2001), the two variables were separated, since it was believed that the generation of images was different from their integration into practice.

Cogan (2006) also indicated that imagery could be used as an energizing or activating force when a gymnast reached a mental block to performing a standing back tuck somersault on the beam: 'Instead of focusing on her legs being bolted to the beam, she should imagine her legs feeling like pistons that could propel her off the beam and through the air' (p. 652).

Not all gymnastic movements are of equal difficulty, and imagery when combined with emotions should be directed to the most demanding elements just prior to mounting an apparatus. In this way, each routine is *segmented* into its key difficulties and mental practice should be directed with appropriate emotion to each of these key skills. More comprehensive imaging would probably be appropriate for the pommel horse or the balance beam, because of their complex and continuous nature.

Suinn (1993) pointed out that imagery and mental practice techniques can be used to achieve a variety of goals, such as enhancing correct responses, simulating competitive environments, and eliminating anxiety or negative thoughts. Mental practice is reported to have beneficial effects on the learning of new skills. Certain factors were identified to mediate the effectiveness of mental practice at different skill levels (Murphy & Jowdy, 1993). Individuals who were better imagers, that is, those who can create clear, real, controlled images, were benefited more by mental practice than their less able counterparts. Moreover, it was suggested that experienced athletes may benefit more from mental practice than novices (Suinn, 1993).

Mahoney and Avener (1977) found the *imagery perspective* to be an important factor having a possible influence on the effectiveness of mental practice. They defined imagery perspective in the following way:

> In external imagery, a person views himself from the perspective of an external observer (much like in home videos). Internal imagery, on the other hand, requires an approximation of the real life phenomenon such that the person actually imagines being inside his/her body and experiencing those sensations which might be expected in the actual situation.
>
> (p. 137)

Imagery outcomes were identified as another mediating factor of mental practice. They suggested that negative imagery, that is, individuals rehearsing a task with a negative outcome, has a debilitating effect on performance. One explanation was that 'negative mental practice affects performance through its impact on dynamic properties of the subjects such as confidence, concentration or motivation' (Murphy & Jowdy, 1993, p. 230).

For example, in rhythmic gymnastics, the verbal mental practice of saying 'Don't drop the club or ball' invariably leads to making these errors. It is like telling someone: 'Don't think of the colour red' – the person *usually* thinks of the colour red. Athletes should incorporate positive imagery into their mental practice, since it is believed to prevent them from focusing on negative images and consequently to maintain consistent performance.

Some have viewed imagery rehearsal as a procedure individuals use to optimally arouse or physiologically activate them for a given performance. However, results of studies attempting to test these ideas have been inconclusive. In some studies, the introduction of arousal in imagery rehearsal did not enhance performance. One explanation for this has been that the arousal included in the imagery increased the activation beyond optimal levels.

One popular skill that has often been cited in sport psychology is that of *self-talk*. From our perspective, this is a sub-component both of imagery, mental practice, focusing and refocusing and is the sub-vocalizing of key words or phrases requiring special attention during the performance of a skill in a routine. For example, during a somersaulting vault, the imagery or mental practice may include a short verbal cue or trigger like: 'Attack the horse and drive back your heels'. In more complex events with multiple skills, the performance should be segmented into automated, or well learned components, with an emphasis on the more difficult or dangerous elements, where self-talk during practice or performance is most critical. For example, for a skilled gymnast, a glide kip on the parallel bars requires little attention or self-talk, while a twisting double somersault requires full attention and vigorous self-talk for greater mental practice, with the sub-verbal cues of 'attack now', 'wrap tightly' or 'stick the landing'.

### 15.6.2 Implications of imagery and mental practice for learning and performing in gymnastics

The most important thing about imagery and mental practice is making the images clear and precise, and then putting them into practice in the most efficient way, which may include the anticipated actions, thoughts, emotions and results. Sylvie Bernier, a Canadian who won the 1984 Los Angeles Olympic gold medal in springboard diving, reported that she felt foolish when she first attempted mental imagery and practice which had been suggested to her by a sport psychologist. But she began to do it, anyway, hidden in her bedroom.

After a few days, when she was on the board, she had a strange revelation that she had already practised this difficult dive in her mind, but now it was much simpler. As she approached the Olympics, she was often doing 80% mental practice and only 20% physical training, which of course, was much easier on her body. She expanded her mental framework, and imagined herself winning the gold medal, then walking up to the medal podium and hearing the playing of 'O Canada', her national anthem, when she was awarded the gold medal! It is truly powerful what the mind can accomplish.

But you must first establish: what is the most effective form of imagery and mental practice for you? Is it best to see yourself, as if on a video, or to see the routine as you would actually perceive it internally in real time. I know that I used the latter technique before going to sleep in high school and my palms would be sweating as I went through each routine.

Self-perceptions of one's gymnastic movements are not always the most accurate measures of what your body is actually doing. Most often, the coaches provide critical feedback of the demands of the key elements of your gymnastics performance. However, with the advances in video technology, it is very simple for gymnasts to observe themselves on their cellphone videos for an accurate performance assessment, which either can be assessed alone or with their coach. This form of external imagery, must then be transformed and internalized into personal imagery, so that the feelings of the movements and the emotions can be rehearsed during mental practice sessions.

The first time that I ever saw a gymnast use this technique was in 1963 at the University of Michigan with twice NCAA parallel bar champion Arno Lascari. Lascari was way ahead of his time. He had someone film every one of his routines over the season with a Super-8 film camera, and he would develop the film every week and study his technique on each event to help both his mental and physical practice.

Eberhard Gienger, the great German champion and a World and Olympic medallist, was observed in a project which documented the competitive behaviours of gymnasts of different genders, nationalities and skill levels (Salmela *et al.*, 1979). Gienger was the captain of the German team and was, by far, the best gymnast on every event. His social and mental preparation procedures were identical for all events in which he participated during the preliminaries, all-around, and event finals. He would actively attend to the individual performances of his colleagues during the first four of six performances and would give them all hugs and social support after each routine. Being the best gymnast on the team, he always performed last. When the fifth gymnast mounted the apparatus, he adjusted his equipment, put on his hand grips and then turned his back to the podium and began his mental preparation. When the fifth gymnast landed, he ignored him, mounted the podium and readied himself for his own performance. It was the perfect mixture of using his social skills as the leader and captain, while giving himself the necessary mental preparation time for his own routine.

Since not all the elements in a gymnastic routine are normally of equal difficulty, it is essential that there are sequencing and priorities established, for it is essential that there is a plan for the next day for the successful execution of the full gymnastic routines. The imagery and mental practice processes should be employed mentally in the same way that we function in our professional or academic lives or sport activities.

For example, the night before I begin working on an article or a book, I focus upon the theme that I want to write about: What makes sense and reads well, or is in need of major revisions, and what is rubbish! Since I am now retired and living in Brazil, I have more liberty and fewer time constraints, so planning my writing and editing takes most of my thinking time. But I also decide upon how much time I should spend on catching up on world events on BBC or CNN, going for a walk, playing with the dogs, talking with my wife Luci or swimming in the pool. The same thinking practices must be incorporated in any serious gymnast's mind: a couple hundred handstand pushups, or looking at Youtube!

It was the same for my imagery and mental practice when I was a junior national level gymnast in Canada. For example, when performing on a concrete or wooden floor in the free exercises, I began with a balance Y scale, which required no attention, and I then did a front handspring front somersault, which required my full attention, given the predictable heel bruises which would occur on these hard surfaces. In the middle of my routine, I included even simpler tumbling, balance and flexibility sequences and then my attention turned to the now simple, round-off, flic flac, back somersault. It is clear with the increased complexity of current routines accomplished on sprung floors, the imagery sequencing processes are much more detailed and the complex skills are increasingly difficult, but the imagery concepts remain the same.

Another related concept in terms of training, imagery and mental practice, is that unsupervised normal athletes spend more practice time on the easy elements that they can easily do. I spent a lot of time doing forward and lateral splits, since I was naturally flexible, but should have devoted more time to the dangerous tricks which terrified me, because of the number and the thinness of the mats and the primitive state of our hand grips, compared to the current ones. While I still love and appreciate my loving high school gymnastics coach Don Cochrane and my inspirational football coach Ivan Livingstone, they were not demanding enough on me. In retrospect, being a tough cookie, I would have run through thicker walls if they had asked me to. I would have responded more to the coaching philosophy of the great Dallas Cowboys' football coach Tom Landry: 'To get men to do what they don't want to do in order to achieve what they want to achieve. That is what coaching is all about' (Irwin, 1993, p. 1).

### *15.6.3 Focusing*

Most mental skills and techniques, including goal-setting, relaxation, activation, imagery and mental practice, require excellent attentional control or focusing abilities. In fact, the ability to consistently attend to the most relevant tasks and environmental stimuli is often referred to in the popular literature as *focusing*, a central aspect of athletic performance (Boutcher, 1993; Nideffer, 1988). Over the years, this construct has been studied from various perspectives, including information-processing and social psychology.

Within the information-processing perspective, attention has been characterized as 'The ability to switch focus from one source of information to another and the amount of information that can be attended to at any one time is limited' (Boutcher, 1993, p. 252). Researchers using this perspective have concentrated their efforts mainly on selective attention, capacity and alertness. Selective attention is believed to play a central role in both the learning and performing of sport skills. It occurs when individuals process certain amounts of information at a particular moment, while other information is screened out or ignored. Research has revealed that selective attention can be voluntary or involuntary, and can take place in a wide variety of behavioural situations.

Attentional capacity is another aspect of focusing that has been investigated. Studies have indicated that there is a limited capacity for processing information at any one time, and that this capacity is even more limited when individuals are engaged in controlled rather than automatic processing. Athletes performing multiple tasks or attempting to focus on more than one source of information could thus experience reduced performance. This often occurs when coaches give too many detailed instructions just before a gymnast's performance. Shriffin (1976) revealed that, although control processing may be dominant in the early stages of learning, it will eventually be replaced by automatic processing, if the skills are to be performed in an effortless and efficient manner.

Arousal is a third aspect of attention that has been examined through an information-processing perspective. Studies have shown that when emotional arousal is increased, focusing fields are reduced and their ability to respond to peripheral stimuli may be decreased (Easterbrook, 1959). Boutcher (1993) reported that this attentional field-narrowing phenomenon may be important to consider in sport performance, since many sport skills are performed in aroused states. Nideffer (1988) reported that when high arousal levels persist, attention may be directed inwardly to dimensions such as fatigue and pain, and the external environment is not considered any more.

Nideffer (1976) demonstrated that individual differences exist in one's ability to use different attentional processes and concluded that individuals possess various styles. It was suggested that the attentional demands of any sport will vary along two dimensions: width (broad or narrow) and direction (internal or external). A broad external focus should used to focus attention on a wide area of the external environment, such as the tasks of a midfielder in soccer, whereas a broad internal focus should be adopted to direct attention internally on various strategies and past experiences, as a coach should during team games. A narrow, external focus was most useful to focus attention on a single aspect of the external environment, as in rifle shooting, while a narrow, internal focus was effective for attending to specific internal images or bodily cues, such as balancing in gymnastics. In the sport of rhythmic gymnastics, narrow internal attention must shift from balance and turning moves to narrow external cues, for catching the object thrown into the air. Research findings on the relationships between attention and performance have incited researchers to develop optimal attentional training programmes for athletes. To help the future development of such programmes, Boutcher (1993) suggested that:

> The precursor to successful attentional control during actual performance may be the establishment of a series of behavioural, physiological, and cognitive cues that optimally prime both body and mind for the ensuing skill.
>
> (p. 262)

Cues or behaviours that lead to optimal attentional states have interested many scholars. In sport, optimal attentional states have often been termed *peak performance* or *flow states* (Csikszentmihalyi, 1975). Flow states have been associated with positive emotions, extremely focused attention and total connection or oneness with the task at hand.

### 15.6.4 Implications of focusing for learning and performing in gymnastics

It has already been mentioned in the previous mental imagery/practice section that it is important to separate into discrete sequences the competitive performance that gymnasts are about to accomplish. They are no longer lying in bed thinking about what they might be doing, they are now focusing on the competitive battlefield and are about to go to war!

In 2004, I met Nelli Kim in an FIG course in Kiev, and I asked her how she finished and who won the all-around championship in the Moscow Games in 1980, since Canada followed the USA in boycotting the Games. She said it was Yelena Davidova from the Soviet Union, with Nadia Comaneci as a close second. She explained that she was in a horse race with Nadia, and on the last event, where she was on the floor exercises, while Nadia was on the uneven bars. Just before Nelli's final tumbling run, she was doing some low-attentional balletic moves in one corner, and she briefly directed her attention to the bars where she saw Nadia fall, and then thought that she was guaranteed the prestigious title. But in Nelli's final tumbling run, which she had successfully accomplished thousands of times, she over-rotated her double back, and landed on her butt. If only she had then known about appropriate focusing, which should have been directed to her final tumbling run, and not upon Nadia!

Hardy Fink, now in charge of the FIG Coaching Academies, related the following to me regarding Boris Shaklin, six times gold medallist in the Olympics for the USSR. Historically, Boris was probably the first gymnast to systematically prepare himself mentally in a consistent fashion, even before mental training was even conceived of:

> I could always recognize him as he prepared to perform a routine. He stood facing away from the apparatus with his arms hanging at his side and his head hung forward and upper back rounded – kind of a position of relaxation, but also ideal for focusing on the task ahead. I guess he probably also had his eyes closed because that seemed natural in that position.
>
> (Personal communication, February 23, 2009)

Nideffer (1976), who was the first to coin the term *focusing*, reported that it was essential to direct one's attention towards something that was important in the sporting environment. It has been demonstrated that gymnasts can focus externally on the crowd, the judges or other competitors, or upon their mental images, or on the state of their own emotions.

The effects of stress on attentional width and external focusing have already been described. I have run three marathons and have watched many more. At the beginning of all marathons, the thousands of amateur runners are all smiling and waving at the applauding people for about the first three kilometres. By the midpoint at around 21 kilometres, their focus narrows onto the shorts of the runner in front of them. Over the last couple of kilometres, their amateur eyes have turned inward, seeing nothing and just hoping that the pain would go away, or even wondering if they could abandon the race!

In gymnastics, the physical stresses are not the same, but the mental ones are. I was at the World Championships in 1987 in Rotterdam, with the Canadian men's team, who were trying to qualify for the Seoul Olympics the next year. Canada was currently in ninth place after the compulsories, with the top 12 teams qualifying. Our youngest gymnast went onto the floor for the warm-up, and it was clear to me from 25 rows away that this lad had lost his mental control. He was not focusing on the apparatus and was wandering around, looking lost. To put it concisely: 'His lights were on, but nobody was home'!

Obviously, the stresses of the moment overcame his mental resources for this situation. Canada fell from ninth to thirteenth place, because of our disastrous team pommel horse performance, and we had not considered a back-up strategy. It was only because Cuba withdrew from the Olympics, for either political or financial reasons, that we eventually qualified. At Seoul, with better technical and mental preparation, we ranked again in ninth place, our best international classification. We had learned to mentally train the team better, how to direct our focus to the most important external and internal cues, and how to avoid the many distractions.

A final example is presented on how our mental training programme helped the Canadian team in the 1988 Seoul Olympics. Since we had extensively trained using the relaxation, imagery and mental practice procedures previously outlined, we borrowed an idea from the mental room or bubble concept, and asked them to create a *team bubble* while competing. All six gymnasts were asked to remain together to be protected from outside influences like other competitors, coaches, the crowd and judges. After their successful performances, many of the gymnasts reported that the team bubble concept really kept the team focused.

### *15.6.5 Refocusing*

Researchers have also used distraction theories to try to explain the relationships between attention and sport performance. They have postulated that individuals lose their focus because certain factors attract their attention to task-irrelevant cues. According to Boutcher (1993), processing task-irrelevant information could explain performance decrements in both competitive and less important sport situations. An unlimited number of factors may cause athletes' attention to be directed towards irrelevant stimuli. Some of these identified sources were the presence of worry, self-awareness, family members, team-mates, coaches, competitors, scores, officials, media, sponsors, close relationships, unrealistic expectations and changes in performance levels. Two of the best predictive OMSAT-3 items for the refocusing scale were: 'Errors generally lead to other errors when I am competing' or 'I think that it is difficult to gain control if I am disturbed during my performance'.

Orlick and Partington (1988) reported upon the refocusing skills of the 1984 Canadian Olympic delegation as follows:

> Those athletes who performed at their highest levels consistently had excellent strategies for getting back on track quickly when things didn't go well, or when faced with distractions. Those who were less consistent appeared to need more work in this area to improve the consistency of their high level performance.
>
> (p. 117)

Because distractions in sport and in life are numerous, researchers have emphasized the importance of developing distraction control or refocusing plans. According to Orlick (2008): 'Refocusing appropriately before, during and after the competition are some of the least practised, but most important skills for high-performance athletes' (p. 49). It is believed that to obtain consistent performance in training and in competition, athletes must regularly develop the skill of distraction control. Refocusing is also consistently the weakest of all of the 12 OMSAT-3 scales, probably because it is neither well understood, nor taught.

Anshel and Payne (2006) introduced two interesting strategies regarding the refocusing dimension, i.e., what to do when things go wrong, using both *approach coping* and *avoidance coping*. Approach coping, much as in Hans Selye's (1974) classic dichotomy of the '*fight* or *flight*' syndrome, is fighting. The difficulty with the fight strategy is that it takes time and requires performance analysis of oneself or of the opponent. The avoidance strategy, or flight, means that stressful, instantaneous events, such as lucky scores by the opponent, unscored points or bad calls from the judges, are ignored or put aside, and one's total focus should remain on what lies ahead, especially in gymnastics. Failure to deal with a distraction in gymnastics on one apparatus can sometimes carry on to performance errors on the next event.

The OMSAT-3 refocusing item that was fifth-best at discriminating between the Canadian international and national athletes (Durand-Bush *et al.*, 2001) was: 'If I start losing, I find it hard to come from behind to win'. Elite athletes in this study had better refocusing skills than their less elite counterparts, and, more specifically, they had the ability to redirect their attention to the task at hand when faced with important distractions to have success in competition. Orlick and Partington (1988) found similar results in their study. They reported that the ability to refocus after distractions varied considerably among elite athletes, compared to the other skills that were assessed through interviews and questionnaires.

### 15.6.6 Implications of refocusing for learning and performing in gymnastics

Unlike in team sports, where a bad call or spectator reactions may cause athletes to misdirect their attention, or focus, from the current situation to inner negative thoughts, in gymnastics it is different. When a gymnast falls from an event, they can only see and feel the whole negative panorama of disturbances of this event, and they often centre upon their emotional reactions, such as anger and frustration. They also can focus upon their thoughts of the future consequences of reducing their final rankings. Also, gymnastics is different from other sports, since gymnasts only have 30 seconds to deal with the error. The most frequent cause of distractions is falling from the apparatus, which is most often from the pommel horse or from the beam. Lower-skilled athletes will just chalk up, and while still in their disturbed emotional and cognitive states, they immediately remount in a worse mental state and then fall off again.

A disconcerting event occurred halfway through the women's all-around competition during the 2000 Olympics in Sydney. An alert Australian gymnast noticed that the vaulting horse, which was specified to be set at a height of 125 cm, had been set 5 cm too low. The officials immediately raised the horse and allowed any gymnast who had already previously vaulted the opportunity to vault again.

It was too late, however, for the Olympic favourite and leader at that time, Svetlana Khorkina from Russia, who had already vaulted and crashed earlier in the competition. Distraught that she had ruined her chances for the Olympic all-around gold, she went to the next event, the uneven bars, and also fell there. Later, when the height error in the horse was discovered, she was told she could redo her vaults, but with her low score on the bars, her all-around hopes were already dashed. This instance brought to light two cognitive mental skill errors she had committed. The first dealt with competition planning: either she or her coach should have taken the responsibility of verifying the height of the horse; the second one was in refocusing, or getting back into a normal competitive state of mind, as has been previously outlined.

One interesting incident regarding refocusing occurred in Tehran, where I spent three months, when feedback was provided to the men's national tae-kwon-do team. When the individual OMSAT-3 profile of Hadi Saei, the 2004 Olympic gold medal winner in tae-kwon-do, was projected on the computer screen, I remarked that his refocusing results were the highest that I had ever recorded, since this OMSAT-3 scale is normally one of the lowest. He and his coach remarked that the OMSAT-3 profile was like '... having a mirror put in front of your face'. Apparently, during the 2004 Athens Olympics semi-final, with three seconds left to go in the match, and being behind by two points, he yelled at and chased his opponent from the mat, who then was disqualified; later on, he went on and won the gold medal. This was an unusual but effective refocusing strategy.

Ken Ravizza, who has worked with women gymnasts in California, outlined a *Six-R* strategy of dealing with adversity in 30 seconds or less:

1 *React*. First of all, curse or swear, but immediately rid yourself of this emotional state, because this anger will never help when you remount. You can quickly call yourself an idiot and then forget about it.
2 *Release*. Try to forget about the fall, or as Orlick (2008) termed it, 'park' the incident, as if putting it outside in your parked car to be dealt with later, or 'change the channel' from a bad movie to a good one.
3 *Review* (if you have time). Try to figure out what went wrong and what you should do when you remount – for example, either repeat the same move or start with the next sequence. Your choice will affect the judges' final evaluation, since correctly repeating the missed element will demonstrate your confidence.
4 *Regroup*. Try to get into your normal pre-competitive state and ensure that you have placed your shoulders back and that your chest or sternum is up.
5 *Ready yourself*. Refocus on the movement cues that you will shortly perform, check your hand grips and take a hard look at the apparatus that you are about to conquer. An extra five seconds is rarely deducted if it looks as if you have a plan.
6 *Relax and go*! Take a deep breath, and exhale slowly and blow away all of the anger or frustration that you had built up and prepare yourself for a positive performance, like the thousands that you have already accomplished in practice and in other competitions.

Some refocusing strategies can be simpler. In the 1990s, at the Canadian National championships, the coach of a top 10 ranked gymnast came up to me and said that his gymnast was upset and performed poorly that day because he was watching our top gymnast, Curtis Hibbert, perform and had felt intimidated. He then asked me, 'What should he do?' I replied, 'Don't watch Curtis – have him focus on the equipment in his handbag when Curtis is performing. When it is his turn to perform, tell him, and then have him approach, and only look at the apparatus.' If this gymnast was a horse, it would be called 'putting on his blinders'. He performed wonderfully and moved up five places! The coach thanked me profusely, but it was just plain common sense.

### *15.6.7 Competition planning*

Researchers have suggested that planning is a very important step in the achievement of peak performance or flow states. Williams (1986) reported:

> Each athlete must to learn how to create consistently at competition time the ideal performance state (thoughts, feelings, bodily responses) typically associated with his or

> her peak performance. Rarely will this occur if pre-competition preparation and competition behaviours are left to chance or to good and bad breaks.
>
> (p. 314)

According to Williams, establishing pre-competition and competition routines not only helps athletes develop a consistent performance approach, it also helps control their arousal levels. It was recommended that athletes organize their internal thoughts, feelings, mental images and the external environment in such a way that they can maximize their feelings of control, and cope with unforeseen events. Developing pre-competition and competition plans is a long process which requires constant evaluation and refinement. Williams indicated that trial-and-error experimentation, combined with consultations with a coach or a mental trainer, may be necessary before athletes can establish their most effective pre-competition and competition routines for achieving optimal performance.

One interesting empirical study demonstrated the importance of competition planning and other mental skills in high level sports. Orlick and Partington (1988) assessed 235 Canadian Olympic athletes' mental readiness through questionnaires and individual interviews. It was found that these elite athletes (a) possessed high levels of commitment; (b) set clear short- and long-term goals; (c) did imagery and simulation training; (d) focused and refocused under distractions; (e) had an established mental training plan that was used and refined throughout the season; and (f) had clearly established mental plans for competition, which included pre-competition and competition mental plans, distraction control plans and post-competitive, constructive, evaluation plans. Orlick and Partington (1988) found that between physical, technical, and mental preparation, mental preparation was the only variable that significantly predicted the athletes' actual Olympic rankings.

Planning for a given competition can begin months, even years, before the actual event. Gymnasts must plan what the elements are that they must consistently master, based upon not only current gymnastic norms but also what types of routines will be performed in the next two, three or four years. A dozen years ago, no one could anticipate the multiple, consecutive tumbling runs, both forwards and backwards with various twists, that are currently being performed. If the plan for the next years includes a single tumbling run with a brief pause for the performance of the next series of simpler skills, the gymnast needs to renew their plan.

In the events that lead up to a selected target competition, a list of necessary equipment should be made up in the kit bag. The night before, all of these items should be in the bag, which should be packed by the gymnasts, and not by their parents. A sense of responsibility must be developed and personal ownership of the gym bag is essential. For instance, has it been planned on packing an extra pair of broken-in hand grips, in case one of the old ones break in competition? Are both sets of competition tops or pants packed, in case there is an on-site change by the coach? Are there extra wrist wraps and adhesive tape, or hand spray such as Tuff Skin, in case of a minor hand rip?

Cogan (2006) and Cogan and Vidmar (2000) elaborated a series of competition planning activities that used competition *simulations* to ready their gymnasts. This occasionally included 'mock meets', where the gymnasts were dressed in their full competition gear, a 'one touch' three minute warm-up, with judges, scores and sometimes even awards. Other coaches I have talked to have included playing in practice, recorded crowd sounds on loudspeakers during a competition, as well as cheering and booing after the events to help gymnasts develop their refocusing skills. Cogan also suggested videotaping the routines of all gymnasts to be used in non-gymnastic sessions for technique corrections by the team and the coaches.

All of the previous mental skills, including goal-setting, relaxation, activation, imagery, mental practice, focusing, and refocusing can be integrated into the pre-competition plan.

Each mental skill element must be practised and not left to chance. Most importantly, the plan should include practising the elements that the gymnasts are having difficulty with, and not only those that they have already mastered.

### *15.6.8 Implications of competition planning for learning and performing in gymnastics*

Gymnasts may have had a specific time of day for mentally rehearsing their sport, but they might not have had a plan of what exactly they were going to rehearse. As a young high school gymnast, I had never heard of mental training. But I have attended some competitions where a gymnast realized that he had forgotten his hand grips because his mother forgot to pack them, but was able to borrow them from another gymnast. Even more serious was the situation recounted to me by a colleague in Canada, where a young member of his ice hockey team forgot to bring his skates to a game!

When I was coaching the men's university team at Laval University, on Thursdays, before the Saturday competition, I would hold a competition simulation or 'mock meet'. The gymnasts were dressed in their full competitive uniforms, would march in, after having a general warm-up, to be greeted by four beautiful women judges who knew nothing about gymnastics. The gymnasts always competed better on Thursdays and the judges gave me sheets of paper that were either blank or had some bogus score on them.

Prior to the 1988 Seoul Olympics, we had a very successful three week training camp which included some unique contingency planning procedures. The national coach and I prepared a list of things which could occur or go wrong at the Olympics and how to deal with them. For example, 'What do you do if a team-mate gets injured during the competition'? Some said that they had to help him. *Wrong answer!* Step over the body of this warrior and move onto the next event, because there are plenty of sport medicine doctors and physiotherapists on site, and they must not lose their focus. This actually did occur. Or another: 'What do you do if a rival head judge tries to ignore you to unsettle you and does not give you the hand signal to begin your routine?' Some said wait until he acknowledges you. *Wrong answer!* They were told just to look at the red light on the judging table and when it was lit, they could begin their routine. Many mind games occur in these important events between both gymnasts and judges and you need to have a plan.

## 15.7 Implications of cognitive skills for performing in gymnastics

Cognitive, or thinking, skills are central to exceptional performance in gymnastics. Gymnasts must be able to imagine the ideal gymnastics movement, imaged either as if watching a perfect video, or as if from their personal perspective of actually performing the skill. This is usually accomplished after doing relaxation methods. They should be able to see and feel the ideal movement patterns, once learned.

However, the mental training on these skills must be carried out both in the gym, while performing, and at home in a rested environment. Gymnasts often tell of having vivid images, with accompanied sweating of the palms and an increased heart rate. These cognitive skills go well beyond actual practice situations, and can be effectively used in competitions.

Refocusing is the least taught mental skill and refers to what to do when something goes wrong in competition, usually from falling from an apparatus or receiving a lower than expected score on a given event. There are a number of strategies outlined in the text, which facilitate regaining of the optimal mental state to continue performing well, and recovering from these negative incidents. In the above sections, some strategies are provided which enable rapid readaptation to the current event, to quickly 'get back on track', resume their normal routine and forget the negative incident and its consequences for the current performance.

Competitive planning includes all of the above mental skills, including the foundation, psychosomatic and cognitive dimensions. This means all of the states which precede competition or may occur during competition, and an assessment during post-competition, both positive and negative, must be planned for.

## 15.8 Conclusion

Competitive planning relates to all of the above mental skills. How does a gymnast feel that they will devote at least 10 years of their lives to excel in their sport? More importantly, how will you invest in achieving excellence, even if it is not in the sport of gymnastics? You cannot go anywhere if you do not have a plan – otherwise, you will end up somewhere you do not want to be!

Gymnasts must plan on what they want to think about, how they want to feel, how they have to prepare themselves, how they organize their equipment, and how they act during both training and competition, on a daily basis. One of the most effective ways is by maintaining a daily list in which are included their goals, intended emotional states, how to control them, their necessary thinking states, and how to react when something goes wrong. A small notebook will suffice, and notes should be taken on how to improve for the next practice, or competition. These simple steps can also be used for the rest of their careers, when they start their next profession.

# 16

# CLOSING THE CIRCLE OF MENTAL SKILLS TRAINING

## Providing mental skills feedback to gymnasts

*John H. Salmela*

Since it is believed that all of the 12 OMSAT-3 scales are interactive, but measure distinct variables (Durand-Bush *et al.*, 2001), this permits more effective interventions in the mental training process. This learning can also occur by trial and error experiences in pressure situations, but this latter process is more haphazard and lengthy. This was evident from my interventions with the Iranian delegation, both coaches and athletes, at the Asian Games that even former world champions still sought out consultations regarding their perceived mental training short-comings, which indicated the on-going importance of having mental training consultants available on-site at important competitions. Apparently, the best in the world still want to become even better. A generic overall view of providing mental skill feedback to athletes will now be elaborated with reference to my most recent interactions with the Iranian delegation in the 2006 Asian Games.

Ericsson's (2007) belief is that skill learning is an ongoing lifelong process, if sufficient deliberate practice is accumulated. Unfortunately, the data on the number of practice hours were not collected on these Iranian champions, but the number of years of training for the women approached the ten-year limit for expert performance, while the men surpassed the women by three years.

The athletes in each discipline were first given an overview of the advantages of the mental skills orientation in relation to fixed mental traits or capacities, such as personality or IQ tests, which are commonly used by clinical psychologists in Iran. The total delegation of athletes was administered the adapted Persian version of the OMSAT-3, either in a classroom or, if there were time constraints, at their residence. All groups received at least one team mental training session along with their coaches from the senior author and his two Iranian associates, which included cognitive–behavioural relaxation, and, when deeply relaxed, this led to the visualization procedures described in the above sections. Additional individual sessions were also scheduled for those athletes who wished to have further consultations, and who were usually the best performers.

All athletes and coaches were shown on a laptop computer screen their individual and team OMSAT-3 profiles in a graphic form and were asked whether they made sense of them. They also received a printed version of their own personal profile for their individual study. Their strengths and weaknesses were pointed out and their perceptions validated the assessments. They were shown that improvements on one scale could affect positive changes in others. For example, they learned that they could take control of their state of stress through the relaxation techniques they were taught, which could positively affect their focusing, refocusing, stress control, and confidence levels. They were also shown that competitive planning could have powerful effects on almost all variables.

However, since the delegation was so large and I did not speak Farsi, my interventions were more limited. Also, the women on the teams, in a follow-up questionnaire, responded more positively than did the men, since (I believed) they appreciated more the attention that was given to them (Salmela *et al.*, 2007). However, possibly due to shyness or cultural norms, I was never asked for personal consultations with them.

Many interventions related to mental skill training were also carried out by my trained associates in 15 disciplines over a three to five month period. These included the administration of the OMSAT-3, the interpretations of individual and team profiles, and subsequent individual and team interventions, based mainly upon relaxation and imagery skills training. Individual consultations based upon the profiles of the selected athletes in nine sports were conducted over the month prior to the Games.

Interestingly enough, an elite athlete in Canada in the Durand-Bush *et al.* (2001) study wrote on his retest questionnaire that:

> After having filled out this questionnaire for the second time, it is very clear to me that the majority of the items were consciously answered in a different way from the first to the second time. I have, therefore, been tremendously influenced by the events (particularly the sporting ones) in which I have participated when answering these questions.

During the initial testing, this subject obtained means of 5.67 and 4.78 on the self-confidence and commitment scales, respectively. When retested, this subject scored higher on the same two scales, and the mean scores had increased to 5.83 and 5.78.

It is important to note that between the two testing sessions this subject won the Canadian Championships in his sport, and it was therefore not surprising that his level of self-confidence and commitment had increased. This reflects the fluctuating nature of developing skills in relation to situational outcomes. It also points out that the simple administration of the OMSAT-3 is an *intervention* in itself. The items revealed some aspects of mental preparation that some athletes perhaps had never considered, or in other cases, reinforced some mental skill dimensions that they had discovered from trial and error, but were never taught.

As an example of the latter, in 1992, on my first visit to Iran, I had a university professor translate the items of the original OMSAT-3 to a group of four women rifle shooters. When I asked them for feedback on the test, most of them replied that they could never believe that their innermost thoughts and feelings regarding shooting had already been written down on paper! This again indicated that these mental skills can be learned by trial and error, but it just takes longer, it occurs at a later stage in their careers, and there may be gaps in knowledge.

For instance, during a graduate level class in Brazil, I had a dozen students complete the OMSAT-3 and a computerized graph of their mental skills profile was generated of what they were doing, thinking and feeling when they were at their best in sport. They then exchanged their profiles with each other and were to do a diagnostic report of their colleague for the next week's class.

During the debriefing the following week, one expert indoor soccer coach showed extremely high levels in his foundation skills, but his cognitive skills were extremely low. When questioned about these low values, he said that coaches in Brazil, when he was a developing athlete, never mentioned imagery, mental practice, focusing, refocusing or competitive planning. This fact alone is a good justification for the teaching and practising of these mental skills at a young age to increase developing players' chances of winning.

It was thus concluded that the OMSAT-3 is more a state- or situation-oriented inventory that will yield different scores when the athletes fill it out depending on the time of the season or the stage of their career in sport. Because it was anticipated that the athletes' responses would vary within the various training and competitive phases that they progressed through in a season, it was suggested that the reliability of the OMSAT-3 scales should not be assessed entirely upon test–retest coefficients. The strengths and weaknesses of athletes' states of mental training should be considered as a reflection of the mental skills and practices over that period of time.

## 16.1 Conclusion

It is scientifically evident that high scores on the three foundation skills are prerequisites for exceptional performance in gymnastics and in many other of life's challenges. If a gymnast does not have clear and, sometimes, public expression of their short- and long-term achievable goals, where do they think that they are going? If they do not have the self-confidence that they can achieve these goals – from within or from their coaches, family or a mental trainer – chances become even slimmer for success. Finally, if they are not committed to investing over years of thinking and training, by themselves, or with their coach or mental trainer, they should get themselves measured for a shovel, and work as a gardener! Fournier *et al.* (2005) demonstrated that most of the selected OMSAT-3 mental skills – with the exception of stress control, which may be another form of mental toughness, developed in their upbringing, or due to genetics – showed improvements after a season-long training with mental skill interventions.

It was concluded in Orlick and Partington's (1988) study that crucial elements of success for the best athletes in the world were (a) total commitment, (b) quality training which included daily goal-setting and imagery training, and (c) quality mental preparation for competition, which entailed developing pre-competition and during-competition focusing and refocusing, and post-competition evaluation plans. Similarly, it was found by Mahoney and his colleagues (1987) that the top level athletes in their study (a) were more confident, (b) were better able to focus before and during competitions, (c) were less anxious, (d) had better internally focused imagery abilities, and (e) were more committed to excelling in the sport, than competitive athletes in the lower ranks. Finally, with a large sample of international Iranian athletes, the total number of OMSAT-3 scale differences were almost identical between Canadian national and international level athletes. However, the scale values were reduced to two – stress control and refocusing –for selected and non-selected international athletes, and then only the one scale of stress control for medallists and non-medallists, which indicated that training and the relative levels of expertise of the athletes showed up in their mental training abilities (Salmela *et al.*, 2009).

Many of the above findings clearly demonstrate both the scientific and practical reasons for the selection of the OMSAT-3 methodology. First of all, the OMSAT-3 provides quantitative data and interpretations via the Internet at a reasonable price. Other instruments are more limited in their scope, and it regroups the theoretical underpinnings of these measurements. From these perspectives, I personally feel that the present comprehensive perspectives from sport psychology research, dedicated to the single sport of gymnastics, has personally allowed the collation and integration of new insights for practical applications, neither of which existed when I edited my first book on gymnastics, more than 30 years ago (Salmela, 1976).

The final mental skills interactions required to close this performance circle will now be addressed. Research in Canada (Durand-Bush, *et al.*, 2001), France (Fournier *et al.*, 2005) and Iran (Salmela *et al.*, 2009) has shown a number of relevant elements which contribute to the understanding of mental skills and exceptional performance in sport. The most important elements are their interactivity, or how learning within one domain influences one or many other mental skills.

First of all, the research on the development of expertise in sport began qualitatively with Bloom (1985), with champion swimmers and tennis players, quantitatively with Ericsson *et al.* (1993) with musicians, with Côté *et al.* (2003) qualitatively in a variety of sports in different age groups, and with Durand-Bush and Salmela (2002) with multiple world and Olympic champions. Based upon his personal experiences with expert sport performers, Orlick (2008) suggested some key elements of mental skills of champions, such as commitment, belief, or self-confidence and commitment. However the links between the interactions of expertise in sport and a variety of mental skills were first demonstrated by Durand-Bush *et al.* (2001) using quantitative measures with the OMSAT-3 and by Durand-Bush and Salmela (2002), with multiple Olympic or World Champions, using interviews.

What was clear from the OMSAT-3 research was that expertise levels were clearly related to almost all of the assessed mental skills between the more experienced international athletes, and national level athletes, who were skilled, but who had not yet competed internationally. Based upon Orlick's assumptions, the foundation skills were prerequisites for excellence.

In the OMSAT-3, the three foundation skills were somewhat modified from Orlick's intuitions. In the Salmela *et al.* (2009) study, it was demonstrated that with the international level Iranian athletes, whether they were selected for the Asian Games or not, or were medallists or not, all showed no significant differences between their foundation skills, as was demonstrated with the Canadian sample. The question is does success in sport increase the levels of goal-setting, self-confidence and commitment, or the reverse? My guess would be that success improves these levels of mental training, especially given the lack of interventions in sport psychology in Iran.

Within the categories of the nine remaining psychosomatic and cognitive skills, several issues remain unclear. While it has been shown that high levels of anxiety narrow the focus or attentional fields of performers (Easterbrook, 1959; Nideffer, 1988), there is very little research to support other interactions that could close the circle of all of the mental skills in sport performance. One obvious place to begin such investigations would be the consideration of the role of relaxation with the other scales, since it was close to discriminating significantly between the selected and non-selected

Iranian athletes. Durand-Bush (1995) showed that relaxation was positively correlated with 10 of the OMSAT-3 variables, with the exception of fear control, while self-confidence was positively correlated with all of the other scales. Interestingly enough, goal setting was positively related to most scales, with the exceptions of stress, fear control and refocusing, while commitment was not positively correlated with stress and fear control. Obviously, stress and fear control, as well as refocusing, require further attention both by researchers and mental trainers.

It also seems obvious that competitive planning, when combined with relaxation and imagery, would also have positive effects upon almost all of the variables, with the exception of fear control (Durand-Bush, 1995). Good coaching from both the conceptual and biomechanical perspectives, given the appropriate audio-visual, material and technical resources, could greatly reduce fear factors in gymnastics.

Much work in research and interventions in gymnastics has been accomplished worldwide using the various sport sciences, but there still remain many questions to be resolved, with perhaps the most in sport psychology and mental training.

# PART III REVIEW QUESTIONS

Q1. Discuss which men's and women's events have similar task demands and others which are completely unique.

Q2. Can gymnasts be evaluated by a single sport science method, and why?

Q3. Why is gymnastics an unnatural sport for human beings?

Q4. What is the most appropriate method of evaluation of a gymnast's potential?

Q5. Discuss the differences between the four models of expertise development in sport.

Q6. Discuss the roles of coaches in a gymnast's development.

Q7. Discuss the roles of parents in a gymnast's development.

Q8. Discuss the most important interactions between the foundation skills.

Q9. Discuss how psychosomatic skills can affect cognitive skills.

Q10. Discuss why you have to plan ahead of training or competition, regarding your goals, emotional states, and cognitive activities.

Q11. Why is it worthwhile to keep updating your note book when you are progressing throughout your gymnastic career, or your future lifelong activities?

# PART IV

# Interaction between physiological, biomechanical and psychological aspects of gymnastic performance

## Introduction and objectives

*Monèm Jemni*

The objective of this part is to give an insight into how physiology, biomechanics and psychology interact with each other in order to contribute to a better understanding of gymnastics performance. We aim to address the coaching processes, the pedagogical and performance analysis as an entire picture.

# 17

# THE PHYSIOLOGIST'S POINT OF VIEW

*Monèm Jemni*

As explained in the introduction of the physiology chapter, the aim of exercise physiology is to understand how human systems work under different exercise conditions and regimes. Humans react differently under different stress levels and in various arousal states. Such conditions have various impacts on hormonal regulation, which affect muscle physiology and force production and changes the neural control of the movement pattern (Mikulas, 1994). Indeed, the impact of a 'stress situation' differs between the contexts of gymnastics training and competition. The way that gymnast interacts with the external environment (coach, team-mates, spectators and even the apparatus) changes according to the situation and subsequently, influences the quality of his or her performance.

The 'environment' in which gymnasts perform is very special because of the specificities of the apparatus. Safety is a major concern in this sport where high risk of injuries is associated with the high acrobatic elements. Equipment engineers have contributed to the immense evolution of gymnastics from the 1970s until now. Coaches, gymnasts, medical staff, physiologists, biomechanists and psychologists interacted with manufacturers, each from their respective points of view, in order, not only to improve the safety of the practitioners, but also to insure a parallel evolution of the equipment to match the increasing difficulty of gymnastics.

The following paragraphs give more evidence of these interactions.

## 17.1 Body composition versus physiology, biomechanics and psychology

Bale & Goodway (1987) analysed the performance variables associated with competitive gymnasts. They showed that male gymnasts generally reach their peak of performance in their early twenties, whereas female gymnasts tend to reach their peak in their mid-to-late teens. Working with different age groups is one of the biggest challenges for a coach. Each age group has different 'group psychology', 'group personality' and also 'group fitness', which might be totally different to the respective individual components. In fact, the 'individual personality' merges into the group's personality to create a 'trend'. In the meantime, some individuals might have stronger roles or status than others in each of the above groupings. The coach has to understand all these components of group psychology while considering the individual variables. There are many coaches who succeeded in working with younger age groups but completely failed with older gymnasts, and vice versa. Shall we end up by accepting common sense that says, 'Oh, this coach is born to work with children!'? In addition, how many times have you heard about conflicts between a coach and a gymnast? How many times have you heard about a very talented gymnast who was a huge success at a junior level but disappeared or burned out after a few years? How many times have you heard about severe injuries in gymnasts? Have you ever tried to understand why?

Working with younger gymnasts has crucial advantages. Young gymnasts are typically lighter and shorter than older gymnasts. This slender physique has biomechanical advantages in performing high risk acrobatic skills common to contemporary gymnastics. It is widely accepted that the short stature and low weight typically observed in elite gymnasts offers a less moment of inertia to the angular momentum dominant in gymnastics performance (Faria & Faria, 1989). Coaches consider the pre-puberty phase as an important period: gymnasts are very receptive to technical learning and strength and power gains during this period. It is also well known that puberty is associated not only with morphological and hormonal transformations, but also with an increase of maximal power (Bedu *et al.*, 1991; Falgairette *et al.*, 1991).

It is very common that high level gymnasts practise for more than 20 hours per week during this period. However, it has been shown that the high level of stress imposed by training and conditioning has an impact on menarche, bone health and growth and development (Courteix *et al.*, 2007; Jemni *et al.*, 2000; Sands *et al.*, 2002; Theodoropoulou *et al.*, 2005) – see Section 6.3. Thus, young children are potentially at greater risk of different types of injuries. It is therefore very important to understand all of these issues when working with young gymnasts. Key elements that can avoid these heath risks are overload and progression, individual response, readiness and periodization. These are, in fact, some of the most important principles of training, which a coach has to fully consider. A very clear and progressive periodization of the training seasons, taking into account all the above, is necessary to avoid burnout. In addition, physiology provides some tools to assess the progress of the gymnasts before raising the load stimuli, taking account of their stage and readiness. Some of these tools might be applied as tests in the laboratory or in the gym.

The previous physiology chapter has demonstrated that gymnasts (mainly females, artistic and rhythmic) are prone to an increased risk of eating disorders and mineral deficiencies, with a remaining issue: if they eat enough, *what* do they eat? (Filaire & Lac, 2002; Jankauskienė & Kardelis, 2005; Lindholm *et al.*, 1995). In addition, many gymnasts and coaches are at risk of extreme behaviours in an attempt to maintain a lean body composition; including extreme dieting, dehydration by the use of diuretics, punishment and food deprivation by the coach if the gymnasts put on some weight, and so on. Evidence has shown that inappropriate diet leads to a decreased performance and an increased risk of injuries and some severe health issues (Benardot, 1999).

In order to prevent such issues, physiology provides tools to monitor gymnasts' health and body composition via standardized tests and blood or saliva analysis. Regular medical/physiological exams are indeed recommended, particularly at high level. It has also been shown that gymnasts are more open with the medical staff than with their coaches this, in fact, might improve prevention, help the diagnosis of certain conditions in order to get treated appropriately and ensure healthy training and practice. Moreover, authors such as Borgen & Corbin (1987) and Rosen & Hough (1988) have suggested making the role models, such as parents and coaches, more sensitive to the pressure towards exaggerated slenderness that the gymnasts experience and encouragement give sensible advice on nutrition. The development of the sport psychologists' role in this context can be a great support in many difficult cases.

## 17.2 Skills design

One of the main areas where biomechanists, coaches and physiologists effectively interact is in the 'invention and design of new skills'. It is thanks to the collaborative work of these three types of specialists that gymnastics has seen an extraordinary expansion of their technical repertoire. There is evidence of work that has been previously performed using computer simulators in order to address the required biomechanical variables, in particular for high aerial acrobatic elements on the high bar, parallel and uneven bars and vault (Hars *et al.*, 2008; Holvoet *et al.*, 2002; Mkaouer *et al.*, 2008; Mkaouer *et al.*, 2005; Sands *et al.*, 2005; Sands *et al.*, 2006a; Sands *et al.*, 2006c; Yeadon *et al.*, 2005). Some of these studies have been published but others were kept secret. Also, some of these simulations have been successful but others did not show any applications (Know *et al.*, 1990; Milev, 1994; Petrov, 1994a, 1994b). A close link with the physiologists is a key point to guarantee the success of a newly 'designed' element. Body composition modelling is very important to consider to guarantee a new successful skill.

## 17.3 Growth and development versus personality

Heredity, maturity, gender, nutrition, rest, sleep, level of fitness, illness/injury, motivation and environmental conditions influence the responses of a gymnast to a training stimulus. Obviously, each gymnast's response is different from that of another. For these reasons, one of the principles of training to fastidiously apply is individualization. A wise coach should detect individual responses and formulate appropriate reactions for each athlete. However, as explained in Section 4.3, there are many cases when coaching a team makes individualizing a practical impossibility. The history of gymnastics provides several examples of clashes between gymnasts and their coaches. To protect privacy, this section will not mention any names. However, it has been shown that most of these issues occur during adolescence and could have been avoided if coaches had a wider knowledge and repertoire of tactics about this sensitive period. Indeed, teenagers undergo various physiological changes which directly affect their moods and psychological well-being. Consulting sport psychologists might resolve major conflicts and situations. Few high level gymnasts have indeed seen their career ending because of a lack of collaboration between coaches, psychologists and physiologists.

# 18

# THE BIOMECHANIST'S POINT OF VIEW

*Patrice Holvoet*

The number of biomechanical studies is continually increasing. Even if many skills can be grouped together relative to common biomechanical principles (Brüggemann, 1994), it is unfortunately impossible to study all of the gymnastic movements performed on the various apparatus.

A primary purpose of biomechanical studies is to contribute to a better understanding of complex human movements. Therefore, descriptive motion analysis of existing skills is a necessary step in the development of meaningful principles that explain gymnastic techniques. Biomechanical investigations of gymnastic skills reveal that movements can be performed in many different ways. Due to the multitude and redundant degrees of freedom of the body's articulations, the segmental movements need to be controlled by perceptual-motor processes. For a better understanding of these strategies, mechanical principles have to be combined with psychomotor knowledge that could explain the spatio-temporal organization and regulation of the movement of the body's segments.

A second purpose is to develop specific principles applicable to the creation of new skills. For example, computer simulations attempt skill creation by trying a skill under different conditions to answer questions such as: what is required to execute complex new airborne dismount figures? But are these computer simulations realistic? Initial computer models are checked to determine the physiological values of a joints' torque and muscular activities to ensure that the simulations represent what can be performed by real gymnasts.

The third purpose consists of the contributions of biomechanical studies for increasing safety through equipment development. For this purpose, examinations of contact conditions with the apparatus (landing on mats, bar gripping, or springboard takeoffs) are conducted to guarantee better absorption of shocks, reduction of force and impulse peaks, and greater stability.

Let us look now at some findings of the biomechanical gymnastics studies and discuss how biomechanics must interact with other scientific disciplines to contribute to a better understanding of gymnastics performance

## 18.1 Floor exercise, trampoline and tumbling

Numerous studies investigate the takeoff requirements for performing acrobatic skills such as double somersaults, with or without twists (King & Yeadon, 2004; Yeadon, 1993a, 1993b, 1993c, 1993d; Yeadon & Mikulcik, 1996). Takeoff velocities, linear and angular momentum, body positions and contributions of the body segments to these kinematic characteristics, which may define the main factors that affect performance. The build-up and the control of body configuration changes are examined by comparisons between techniques actually used by gymnasts and computer simulations. These analyses indicate what body orientations enhance performance, and what are the segment motions reponsible to adjust and control rotations during the flight.

Even if it is difficult to convey the results of these studies down to the coaches and gymnasts, they are an interesting starting point for investigations on the function of the perceptual systems (such as visual and vestibular systems), which control body orientation and angular velocities during complex acrobatic skills.

For example, the regulation of the body's moment of inertia during aerial skills is of crucial importance for a safe landing. Indeed, the gymnast must first modulate the angular momentum to execute the required degree of body rotation and must secondly, reach a body configuration that permits absorption of the landing shock, while maintaining balance. One of the main characteristics of the gymnast's expertise is the stability of aerial movements. When performing complex aerial skills, high level gymnasts are able to reproduce appropriate body configurations with great consistency.

Recent psychomotor studies explain this stability as the result of a control strategy which requires that the gymnast pick up information concerning future body orientations relative to the ground, distance to the ground, and time remaining before ground contact (Bardy & Laurent, 1998). This information is used to initiate body extensions for decelerating the angular velocity and regulating the angular displacement of the body at landing. These studies, based on protocols of trials 'with and without vision' executed by novices and experts, indicate that the body configuration is visually controlled during flight by the experts, but not by the novices (Davlin *et al.*, 2001a, 2001b, 2002, 2004).

## 18.2 Vault

Many studies have examined the correlations between kinematic springboard parameters, parameters while in contact with the horse and landing parameters with the scores given by judges, in vaults such as the handspring, the handspring with twist or somersault, and the Yurchenko vault (King & Yeadon, 2005; Koh & Jennings, 2007; Takei, 1998).

It has been demonstrated that the horizontal running velocity and the linear and angular takeoff parameters are very important factors to optimize and to realize effective performance. Simulation models confirm that having appropriate initial kinematics at touchdown are essential and that the use of shoulder torques during horse contact plays a minor role in vaulting performance. Consequently, the meaningful information for coaches and gymnasts to consider is that it is very important for enhancing vault performance to focus on generating high body running velocity and a high level of angular momentum during the takeoff from the springboard.

Even if these studies become more realistic, some important physiological factors are needed to improve the biomechanical models. For example, individual values of stiffness and damping governing the visco-elastic characteristics of joints' elasticity and realistic activation time histories of joints torques are required to determine the ability of joints to resist the compression action during springboard and horse contact.

## 18.3 High bar and uneven bars

The mechanics of giant swings and the takeoff requirements of dismounts and release–regrasp techniques are the most commonly examined performance factors (Brüggemann, 1994;

Holvoet *et al.*, 2002a; Holvoet *et al.*, 2002b). Kinematic, kinetic, energetic and electromyographic data are reported to identify differences, to establish profiles for different techniques and to classify techniques relative to common principles (Arampatzis & Brüggemann, 1998). Even if the construction and design of these types of apparatus are different, similarities between the mechanics of the horizontal bar and uneven bars are established in takeoff conditions or dismounts. Mechanical differences between these two types of apparatus result more from the 'beat' or 'tap' action through giant swings (Sheets, 2008). Differences in the anthropometric and physiological characteristics between female and male gymnasts should be taken into consideration when explaining the specific techniques used to perform similar skills.

Biomechanical studies develop optimization processes using computer simulation models (Hiley & Yeadon, 2008). These models become more and more complex. The bar is often represented as a spring and the gymnast is considered as a three-dimensional articulated system. Shoulder stiffness and damping are also taken into consideration. Comparisons between experimental and simulation performances are conducted to verify how the model reproduces the movement and what parameters can be optimized to minimize joint torques to maximize performance (Begon *et al.*, 2008). As has been seen in the preceding section relative to the vault, there is a need for the addition of individual physiological values for the strengths of joints and muscular activities in validating these optimization processes.

## 18.4 Rings and parallel bars

Biomechanical studies on these types of apparatus are less extensive than those on the horizontal bar. Kinematic analysis and computer simulations have been conducted on the feldge, backward long swing and dismounts on rings (Yeadon & Brewin, 2003). On the parallel bars, two-dimensional models have been developed for making predictions on the swing to handstand and for dismounts, and for analysing the dynamics of the bar acting at the hand grip (Linge *et al.*, 2006). Forces exerted during static movements such as the handstand were also analysed on these two apparatuses (Prassas, 1988).

Because of the important part that strength elements play in performance on these apparatuses, development of the upper body muscular strength and organization of progression in strength training programmes are essential aspects of maximizing technique changes.

## 18.5 Other apparatus

The pommel horse and balance beam are often considered very difficult apparatus because the loss of balance and falls are the primary causes of large score deductions during competitions. Biomechanical studies on the pommel horse are limited to some case studies. Because circling is an essential prerequisite of performance on this apparatus, the control of angular velocity is analysed when performing basic circles, Thomas and Magyar elements (Baudry *et al.*, 2005). Dismounts performed by women gymnasts on the balance beam are the main skills that are examined by the biomechanical studies (Brown *et al.*, 1996). There is a lack of information concerning the nervous control mechanisms that can regulate the posture and the balance involved in the execution of both static and dynamic skills performed on these apparatus. Biomechanical analyses should be combined with neurological and psychological information to better understand a gymnasts' balance ability.

## 18.6 Studies on safety and equipment development

Because gymnastic activities involve repeated large weight bearing impacts on the hands and feet, biomechanical procedures can estimate the internal loads exerted at joints during critical phases of movement (Davidson *et al.*, 2005; Mills *et al.*, 2006). For example, three times the body weight tensile force is exerted at the wrist and shoulder joints when performing the 'beat' swing during a backward giant swing before a release–regrasp skill on the horizontal bar. When performing a simple forward somersault on the floor, 19 times bodyweight compressive force is applied at the ankles. Studies of these types provide useful data for improving the gymnastic training process and for protecting gymnasts from excessive muscular overload. These studies must be associated with medical knowledge of sports injuries to be effective in injury prevention.

It has been seen that an important source of force used in performing many gymnastic skills is the elasticity of apparatus such as the springboards, trampolines, horizontal bars, and asymmetrical or parallel bars. The elasticity of a material is a characteristic which represents a measure of how this material will reform after it has been stretched, bent or compressed. Great efforts have been made to develop models to understand the dynamics of springboard, trampoline and apparatus. A mass-spring system is often used to represent the interactions between the gymnast and apparatus. Stiffness and damping properties of the apparatus were analysed in bouncing or stretching experiments in which given loads are exerted on the materials. In this way, foot contact and hand grasp conditions can be examined during critical phases such as landing. These biomechanical studies are used in several other disciplines whichinclude industrial engineering and manufacturing, research in material and equipment, and innovations in design. For example, there have been many innovations in the manufacture of anti-slide textures that guarantee large safety contact zones on the gymnastic apparatus surfaces.

So improvements in the area of gymnastics depend not only on a better understanding of skill execution, but also on technical progress and innovations in the manufacture of material and equipment.

# 19

# THE PSYCHOLOGIST'S POINT OF VIEW

*John H. Salmela*

## 19.1 Mental skills interactions

This section discusses how the various mental skills interact with exercise physiology knowledge and biomechanical principles for both learning and performing. It is scientifically evident that high scores on the three foundation skills are prerequisites for exceptional performance in gymnastics and in many other of life's challenges. If a gymnast does not have clear expressions of their short- and long-term, achievable goals, where they do think that they are going? If they do not have the self-confidence that they can achieve these goals, either from within, or from their coaches, family or a mental trainer, chances become even slimmer for success. They must be committed to investing over years of thinking and training, by themselves, or with their coach or mental trainer. Fournier *et al.* (2005) demonstrated that most of the selected OMSAT-3 mental skills – with the exception of stress control, which may be another form of mental toughness, whether developed during upbringing or due to genetics – showed improvements after a season-long training with mental skill interventions.

It was concluded in Orlick & Partington's (1988) study that crucial elements of success for the best athletes in the world were (a) total commitment, (b) quality training, which included daily goal-setting and imagery training, and (c) quality mental preparation for competition, which entailed developing pre-competition planning, competition focusing and refocusing, and post-competition evaluation. Similarly, it was found by Mahoney (1989) that the top level athletes in their study were more confident, were better able to focus before and during competitions, were less anxious, had better internally-focused imagery abilities, and were more committed to excelling in their sport than were competitive athletes in the lower ranks. Finally, with a large sample of international Iranian athletes, the total number of OMSAT-3 scale differences were almost identical to those of Canadian national and international level athletes. However, the scale values were reduced to two, stress control and refocusing, for selected and non-selected international athletes, and then to only one, stress control, for medallists and non-medallists. This indicated that training and the relative levels of expertise of the athletes showed up in their mental training abilities (Salmela *et al.*, 2009).

Many of the above findings clearly demonstrated both the scientific and practical reasons for the selection of the OMSAT-3 methodology. First of all, the OMSAT-3 provides quantitative data and interpretations via the Internet at a reasonable price. Other instruments are more limited in their scope, and the OMSAT-3 regroups the theoretical underpinnings of these measurements. I feel that the present comprehensive perspectives from sport psychology research dedicated to the single sport of gymnastics allows the collation and integration of new insights for practical applications, neither of which existed when I edited my first book on gymnastics, more than 30 years ago (Salmela, 1976). The final mental skills interactions required to close this performance circle will now be addressed. Research in Canada (Durand-Bush *et al.*, 2001), France (Fournier *et al.*, 2005) and Iran (Salmela *et al.*, 2009) have shown a number of relevant elements which contribute to the understanding of mental skills and exceptional performance in sport. The most important elements are their interactivity, or how learning within one domain influences one or many other mental skills.

First of all, the research on the development of expertise in sport began qualitatively with Bloom (1985) studying champion swimmers and tennis players, quantitatively with Ericsson *et al.* (1993) studying musicians, qualitatively with Côté *et al.* (2003) in a variety of sports in different age groups, and with Durand-Bush & Salmela (2002) studying multiple world and Olympic champions. Based upon his personal experiences with expert sport performers, Orlick (2008) suggested some key elements of mental skills and demonstrated that variables such as commitment, belief, and self-confidence were essential. However, the links between the interactions of expertise in sport and a variety of mental skills were first demonstrated by Durand-Bush *et al.* (2001), using quantitative measures with the OMSAT-3, and by Durand-Bush & Salmela (2002), with multiple Olympic or World champions, using interviews.

What was clear from the OMSAT-3 studies was that expertise levels were clearly related to almost all of the assessed mental skills between the more experienced international athletes, and national level athletes, who had not yet competed internationally. Based upon Orlick's assumptions, the foundation skills were prerequisites for excellence. In the OMSAT-3, the three foundation skills were somewhat modified from Orlick's intuitions. In the Salmela *et al.* (2009) studies, it was demonstrated that with the international level Iranian athletes, whether they were selected for the Asian Games or not, or were medallists or not, all showed no significant differences between their foundation skills, as was demonstrated with the Canadian sample. The question is: does success in sport increase the levels of goal-setting, self-confidence and commitment? My guess – especially given the lack of interventions in sport psychology in Iran – would be that success improves these levels of mental training.

Within the categories of the nine remaining psychosomatic and cognitive skills, several issues remain unclear. While it has been shown that high levels of anxiety narrow the focus, or attentional fields, of performers (Easterbrook, 1959; Nideffer, 1988), there is very little research on other interactions that could close the circle of all of the mental skills in sport performance. One obvious place to begin such investigations would be the consideration of the role of relaxation, since it was close to showing significant differences between the selected and non-selected Iranian athletes. Durand-Bush (1995) showed that relaxation was positively correlated with ten of the OMSAT-3 variables, but not fear control, while self-confidence was positively correlated with all of the other scales. Interestingly enough, goal-setting was positively related to most scales, with the exceptions of stress and fear control and refocusing, while commitment was positively correlated with all the scales except stress and fear control. Obviously, stress and fear control, as well as refocusing, require further attention both by researchers and mental trainers.

It also seems obvious that competition planning, when combined with relaxation and imagery, would also have positive effects upon almost all of the variables, with the exception of fear control (Durand-Bush, 1995). Good coaching from both the conceptual and biomechanical perspectives, given appropriate audio-visual, material and technical resources, could greatly reduce fear factors in gymnastics.

## 19.2 Foundation skills and exercise physiology

There is no question that the physiological and morphological status of gymnasts from 9 to 14 has a strong, determining influence on the performance of young gymnasts (Régnier & Salmela, 1987). If you are strong, fast, and flexible, this will enhance your goal-setting, self-confidence and commitment to gymnastics. There have been some exceptions to the rule, such as Eberhart Gienger who was quite tall, compared to his team-mates, but obviously, if you are powerful, fit and strong compared to your cohorts, your foundation skills will be high.

Usually in the western world gymnastics coaches do not have specialized degrees in exercise physiology, but have degrees either in physical education or in the sport sciences, and know basic principles of strength, speed, power and flexibility training. They are also able to quickly evaluate whether their young gymnasts have set their training goals, are committed to them, and have developed the necessary levels of self-confidence. This usually results in greater gymnastic skill levels, because of these physiological endeavours on the above physiological dimensions.

However, there are also downsides to commitment from a physiological perspective, and which result in breakdowns, especially in young girls. Girls need to perform earlier in their life cycle than do boys, principally because they have to be able to perform before puberty to aid in the spotting of complex skills, while men must wait till post-puberty to develop sufficient strength to perform, for example, on the rings. Dr Monèm Jemni, editor of this book and a researcher at the University of Greenwich in the UK, has extensive data on this subject (Jemni *et al.*, 2001; Jemni & Robin, 2005; Sands *et al.*, 2000).

For example, as an athlete, both in gymnastics and Canadian football, I remember well my coach having the boys' team before practice do five sets of 10, 20 and 50 yard sprints, back and forth. It killed me. Then I looked down the field and saw two of my idols, Don Clark and George Dixon, who played for the Montreal Alouettes, coming to watch my coach Ivan Livingstone's practice. Ivan called out to me: 'John, you are the only guy who always tries hard in every practice, and is in shape. But, he forgot to bring the bag of footballs! Can you run down to the clubhouse, and bring them back?' I was physiologically exhausted, but sprinted again across two 100 yard football fields with my professional heroes watching me. I refused to fail! Once I got there, I lay down on the clubhouse floor for two minutes to partially recover, and then sprinted back another 200 plus yards, with a bag of more than ten balls. Ivan reminded me of this 20 years later, but I would rather die in front of my idols!

## 19.3 Foundation skills and biomechanics

Bill Sands, a biomechanist on the United States Olympic Committee, was reported on in a recent article, speaking on how to become a champion, based upon physiological and psychological training (Sands, 2003). I had the privilege to work with him for a short time, both in Canada and in Qatar, during seminars. He has a profound knowledge of sport science and gymnastics, and was named as the team coach for the US women's team, which unfortunately, like Canada, boycotted the 1980 Moscow Games because of the Russian attack on Afghanistan, which is now quite ironic. He has keen insights on the interrelationships between the sport sciences and the sport of gymnastics.

I remember, when I was doing my doctorate at the University of Alberta, that the gymnastics coach Tanaka told me that my handstand was too 'archy', or curved, even though I was 25 at that time! He made me lie down on the floor first in a totally stretched and then in a totally contracted position, while two other gymnasts picked up my stiff body, and placed me in the perfect handstand position. Upon graduating, I used the same biomechanical techniques in my university teaching, and, of course, the students' goal-setting, self-confidence and commitment increased. They were then able to learn more complex skills by maintaining this position in both static and dynamic skills, and were thrilled with their mechanical control of their bodies.

## 19.4 Psychosomatic skills and exercise physiology

The psychosomatic skills are probably the most affected by exercise physiology, especially those regarding relaxation and activation. Exercise has profound effects, especially once it has terminated. There is a great feeling of relaxation and relief after an extensive period of intensive effort. This is the moment for deep relaxation, which often enhances mental training and rehearsal of the day's activities and can be combined with imagery and mental training for the next day's training or competition. The combination of the recovery from extreme exertion and its positive effects are ideal for mental training, such as stress control.

On the other side of the coin is that activation activities, such as brief sessions of exertion, even screaming out your key words, usually just in your mind, may pump you up for better performance for certain events, such as floor or vault. Normally, gymnasts do not need to be psychologically activated, especially young girls, since they have proud parents who have invested a great deal of money and have driven them across the country, or even abroad.

In Sabae, Japan, in 1995, at the World Championships for the 1996 Olympics, the whole team and the technical team delegation, lived on the same floor in a hotel. Although I was forbidden to work with the women's team, many of the girls would come up to me and talk, without their coaches. The men were 6 to 10 years older, and the girls often wanted to know why they were so relaxed and joked around, while the girls felt so stressed? The answer seemed so obvious: you are much younger, you worked rather than played at gymnastics, and the men were friends with their coaches, rather than being dictated to by them.

## 19.5 Psychosomatic skills and biomechanics

From my experience, biomechanical knowledge for gymnasts is most useful for the controlling of stress and fear. I have previously described how, at a training camp with teenage gymnasts, who were potential Olympians for the Canadian team in five years, our coaches explained the relatively new move on the horizontal bar, the Tkatchev. For this the gymnast had to perform a backward giant swing and, before the three-quarter point in the giant, had to arch, blindly counter-rotate over the steel bar and regrasp it. The biomechanically trained coaches went through how the proper execution of 'timing techniques', using spotting mats to cover the bar to prevent back injuries, and of deep foam landing pits which would minimize any possibility of injuries. I previously reported how one courageous gymnast went for it, hit the mat on the bar and landed in the deep foam pit, and came up laughing. After that, all 14 gymnasts attempted this element, many times within the next 20 minutes. As described above, I spent several hours talking to the young gymnasts about real and irrational fears, as outlined by Feigley (1987), and they found out that this was, in part, an irrational fear, when using good coaching and some biomechanical and appropriate spotting principles.

## 19.6 Cognitive skills and exercise physiology

In relation to the OMSAT-3, the exercise physiology effects on cognitive skills have already been discussed in the foundation and psychosomatic skills sections regarding the gymnasts' general perceptions of their potential to succeed in the sport in terms of their goals, self-confidence and commitment. Obviously, a non-muscular, fat, slow, and inflexible person will not succeed in gymnastics.

However, if they can learn to relax or physiologically activate themselves, either physically or through mental processes, they may have a chance of greater achievements in gymnastics. However, the psychological dimensions drive gymnasts to work hard, do strength and flexibility training, set goals, keep self-confidence and remain committed, usually through the influences of family, peers, and good coaches.

## 19.7 Cognitive skills and biomechanics

The science of biomechanics, from my perspective, has provided me with the greatest insights on why I was neither a great gymnast nor a university coach, and much of this I have learned in my later years, while participating as a spectator with revolutionary coaches such as the national coach of Canada, Iarov, at international FIG courses in different parts of the world. Yuri was the first coach to have a gymnast, Valeri Luikin, perform a triple back somersault on the floor exercises, and it has since been done only by seven other gymnasts. He could do a double back after just a back handspring! These techniques for tumbling must be mastered within mental practice and imagery skills, since they appear to be much quicker than those traditionally taught.

I also observed unbelievable learning curves within men's support skills on the pommel horse and on the parallel bars. From the periods before the fifties until a decade ago, men gymnasts were taught to swing with their shoulder structures contracted, which elevated and contracted these muscles, and thus limited their degrees of freedom regarding lateral movements in the transverse planes. The upper torso would remain stiffer, and thus limit flexible movements. Iarov showed me and a team of developing Moroccan gymnasts, how the relaxation of their shoulder structures freed up their torsos and allowed them to improve their circles on the pommel horse by 50% within a couple of days.

I also learned within one to two days from Iarov how to teach the complex Diamadov skill on the parallel bars, which required a 180 degree turn on one arm to a handstand, a skill that usually takes months to learn. But the gymnasts were also instructed to imagine the execution of the movement in the evening, and to keep in mind the coach's innovative methods.

Much work in research and many innovations in gymnastics have been accomplished worldwide using the various sport sciences, such as mental training, exercise physiology and biomechanics, but there still remain many questions to be resolved, with perhaps the greatest number in mental training.

# PART IV REVIEW QUESTIONS

Q1. Summarize the point of view of the physiologists on how they interact with the biomechanists and the psychologists in order to enhance gymnastics performance.

Q2. Summarize the point of view of the biomechanists on how they interact with the physiologists and the psychologists in order to enhance gymnastics performance.

Q3. Summarize the point of view of the physiologists on how they interact with the psychologists and the biomechanists in order to enhance gymnastics performance.

Q4. Analyse a real example taken from gymnastics performance where physiology, biomechanics and psychology interact within each other.

Q5. As a coach, how could you guarantee a long lasting career for your gymnasts, taking into consideration physiology, biomechanics and psychology?

Q6. Analyse the interaction of physiology, biomechanics and psychology in the context of skills design.

# REFERENCES


Abernethy, B., Wann, J., & Parks, S. (1998). Training perceptual-motor skills for sport. In B. Elliott (ed.), *Training in Sport* (pp. 1–68). New York, NY: John Wiley & Sons.

Abraham, N., Carty, R., DuFour, D., & Pincus, M. (2006). Clinical enzymology. In R. McPherson & M. Pincus (eds), *Henry's Clinical Diagnosis and Management by Laboratory Methods* (Chap. 20). Philadelphia, PA: Saunders Elsevier.

Alter, M. J. (2004). *Science of Flexibility*. Champaign, IL: Human Kinetics.

Anderson, J. L. (2007). ST segment elevation acute myocardial infarction and complications of myocardial infarction. In L. Goldman & D. Ausiello (eds), *Cecil Medicine* (Chap. 72). Philadelphia, PA: Saunders Elsevier.

Anshel, M. H. (1990). *Sport Psychology: From Theory to Practice*. Scottsdale, AZ: Gorsuch Scarisbrick.

Anshel, M. H., & Payne, J. M. (2006). Application of sport psychology for optimal performance in the martial arts. In J. Dosil (ed.), *The Sport Psychologist's Handbook*. (pp. 353–374). Chichester, UK: Wiley.

Arampatzis, A., & Brüggemann, G. P. (1998). A mathematical high bar–human body model for analysing and interpreting mechanical–energetic processes on the high bar. *Journal of Biomechanics*, 31, 1083–1092.

Arce, J., Haupt, H. A., Irwin, K. D., Ohle, J., Palmieri, J., & Siff, M. (1990). Training variation. *National Strength and Conditioning Association Journal*, 12(4), 14–24.

Argeitaki, P., Katsikas, C., Chondros, K., Diafas, V., & Dessypris, A. (2009). Comparison of enzyme changes in long distance track athletes, football players, cyclists and sprinters. *J Sport and Science*, 1, 120–131.

Arkaev, L. I., & Suchilin, N. G. (2004). *Gymnastics: How to Create Champions*. Oxford, UK: Meyer & Meyer.

Bale, P., & Goodway, J. (1987). The anthropometric and performance variables of elite and recreational female gymnasts. *N Zealand J Sports Med*, 63–66.

Banister, E. W. (1991). Modeling elite athletic performance. In H. A. W. J. Duncan MacDougall, & H. J. Green (eds), *Physiological Testing of the High-Performance Athlete* (pp. 403–424). Champaign, IL: Human Kinetics.

Barantsev, S. A. (1985). Do gymnasts need to develop aerobic capacity? *Soviet Sports Review*, 25(1), 20–22.

Bardy, B. G., & Laurent, M. (1998). How is body orientation controlled during somersaulting? *Journal of Experimental Psychology*, 24(3), 963–977.

Barlett, H. L., Mance, M. J., & Buskirk, E. R. (1984). Body composition and expiratory reserve volume in female gymnasts and runners. *Med Sci Sports Exerc*, 16(3), 311–315.

Barohn, R. J. (2007). Muscle diseases. In L. Goldman & D. Ausiello (eds), *Cecil Medicine*. Philadelphia, PA: Saunders Elsevier.

Bar-Or, O. (1984). The growth and development of children's physiologic and perceptional responses to exercise. In J. Ilmarinen & I. V–lim–ki (eds), *Children and Sport* (pp. 3–17). Berlin: Heidelbergh E.

Bar-Or, O. (1987). The Wingate anaerobic test: An update on methodology, reliability, and validity. *Sports Medicine*, 4, 381–394.

Baudry, L., Leroy, D., & Chollet, D. (2005). The circle performed on a pommel horse in gymnastics: The critical role of double support phase. *Gait & Posture*, 21(Suppl 1), S34.

Baumeister, R. F. (1984). Choking under pressure: Self-consciousness and paradoxical effects of incentives on skillful performance. *Journal of Personality and Social Psychology*, 46, 610–620.

Beaudin, P. A. (1978). Prédictions de la performance en gymnastique au moyen de l'analyse d'une sélection de variables physiques, physiologiques et anthropométriques. Thesis, McGill University, Montreal.

Bedu, M., Fellmann, N., Spielvogel, H., Falgairette, G., Van Praagh, E., & Coudert, J. (1991). Force–velocity and 30s Wingate tests in boys at high and low altitudes. *J Appl Physiol*, 70, 1031–1037.

Begon, M., Wieber, P.-B., & Yeadon, M. R. (2008). Kinematics estimation of straddled movements on high bar from a limited number of skin markers using a chain model. *Journal of Biomechanics*, 41, 581–586.

Behm, D. G. (1995). Neuromuscular implications and applications of resistance training. *Journal of Strength and Conditioning Research*, 9(4), 264–274.

Beilock, S. L., & Gonso, S. (2008). Putting in the mind versus putting on the green: Expertise, performance time and the linking of imagery and action. *Journal of Experimental Psychology*, 61 (6), 920–932.

Benardot, D. (1999). Nutrition for gymnasts. In N. T. Marshall (ed.), *The Athlete Wellness Book* (pp. 1–28). Indianapolis, IN: USA Gymnastics.

Benardot, D., & Czerwinski, C. (1991). Selected body composition and growth measures of junior elite gymnasts. *J Am Diet Assoc*, 91(1), 29–33.

Benardot, D., Schwarz, M., & Weitzenfeld, D. (1989). Nutrient intake in young highly competitive gymnasts. *Journal of the American Dietetic Association*, 89(3), 401–403.

Bergh, U. (1980). Entraînement de la puissance aérobie. In P. O. Astrand & K. Rodahl (eds), *Précis de physiologie de l'exercice musculaire* (2nd ed., pp. 303–308). Paris: Masson.

Bernier, M., & Fournier, J. (2007). Mental skill evaluations of French elite athletes. In Y. Theodorakis, M. Gourdas, & A. Papaionnou (eds), *Sport And Exercise Psychology: Bridges Between Disciplines and Cultures* (pp. 89–91). Thessaloniki: University of Thessaly.

Blochin, I. P. (1965). Energeticeskaja (gazoobmenu) upraznenij sportivnoj gimnastiki u muzsin. *Teor Prakt Fiz Kult*, 28, 32.

Bloom, B. S. (1985). *Developing Talent in Young People*. New York: Ballantyne.

Bobbert, M. F. (1990). Drop jumping as a training method for jumping ability. *Sports Med*, 9(1), 7–22.

Bobbert, M. F., & Van Ingen Schenau, G. J. (1988). Coordination in vertical jumping. *J Biomech*, 21(3), 249–262.

Boisseau, N., Persaud, C., Jackson, A. A., & Poortmans, J. R. (2005). Training does not affect protein turnover in pre- and early pubertal female gymnasts. *Eur J Appl Physiol Occup Physiol*, 94(3), 262–267.

Bompa, T. O., & Haff, G. G. (2009). *Periodization*. Champaign, IL: Human Kinetics.

Bondarchuk, A. P. (2007). *Transfer of Training in Sports*. MI, USA: Ultimate Athlete Concepts.

Borgen, J. S., & Corbin, C. B. (1987). Eating disorders among female athletes. *Physician Sportsmed*, 15(2), 89–95.

Bosco, C., Luhtanen, P., & Komi, P. V. (1983). A simple method for measurement of mechanical power in jumping. *Eur J Appl Physiol Occup Physiol*, 50(2), 273–282.

Bouchard, C., Dionne, F. T., Simoneau, J. A., & Boulay, M. R. (1992). Genetics of aerobic and anaerobic performances. *Exer Sport Sci Rev*, 20, 27–58.

Boutcher, S. H. (1993). Attention and athletic performance: An integrated approach. In T. S. Horn (ed.), *Advances In Sport Psychology* (pp. 251–265). Champaign, IL: Human Kinetics.

Brooks, T. (2003). Women's collegiate gymnastics: A multifactorial approach to training and conditioning. *Strength and Cond J*, 25(2), 23–37.

Brotherhood, J. R. (1984). Nutrition and sports performance. *Sports Med*, 1, 350–389.

Brown, E. W., Witten, W. A., Weise, M. J., Espinoza, D., Wisner, D. M., Learman, J., & Wilson, D. J. (1996). Attenuation of ground reaction forces in salto dismounts from the balance beam. In J. M. C. S. Abrantes (ed.), *Proceedings of the XIVth International Symposium on Biomechanics in Sports* (pp. 336–338). Lisbon, Portugal: Edições FMH Universidade Técnica de Lisboa.

Brüggemann, G. P. (1994). Biomechanics of gymnastic techniques. *Sport Science Review*, 3(2), 79–120.

Burton, D. (1993). Goal setting in sport. In R. N. Singer, M. Murphey, & L. K. Tennant (eds), *Handbook of Research on Sport Psychology* (pp. 467–491). New York: Macmillan.

Burton, D. (1988). Do anxious swimmers swim slower? Reexamining the elusive anxiety–performance relationship. *Journal of Sport and Exercise*, 10, 45–61.

Butt, A. A., Michaels, S., Greer, D., Clark, R., Kissinger, P., & Martin, D. H. (2002). Serum LDH level as a clue to the diagnosis of histoplasmosis. *The AIDS Reader*, 12(7), 317–321.

Carlock, J. M., Smith, S. L., Hartman, M. J., Morris, R. T., Ciroslan, D. A., Pierce, K. C., *et al.* (2004). The relationship between vertical jump power estimates and weightlifting ability: A field-test approach. *J Strength Cond Res*, 18(3), 534–539.

Cassell, C., Benedict, M., & Specker, B. (1996). Bone mineral density in elite 7- to 9-yr-old female gymnasts and swimmers. *Medicine & Science in Sports & Exercise*, 28(10), 1243–1246.

Chen, J. D., Wang, J. F., Li, K. J., Zhao, Y. W., Wang, S. W., Jiao, Y., *et al.* (1989). Nutritional problems and measures in elite and amateur athletes. *Am J Clin Nutr*, 49(5 Suppl), 1084–1089.

Christina, R. W., & Davis, G. (1990). Diving skill progressions: Part 1 Principles of teaching skill progressions. In J. L. Gabriel & G. S. George (eds), *U.S. Diving Safety Manual* (pp. 89–103). Indianapolis, IN: U.S. Diving Publications.

Claessens, A. L., Lefevre, J., Beunen, G., & Malina, R. M. (1999). The contribution of anthropometric characteristics to performance scores in elite female gymnasts. *Journal of Sports Medicine and Physical Fitness*, 39(4), 355–360.

Claessens, A. L., Malina, R. M., Lefevre, J., Beunen, G., Stijnen, V., Maes, H., *et al.* (1992). Growth and menarcheal status of elite female gymnasts. / Croissance et menstruation des gymnastes d'elite feminines. *Medicine & Science in Sports & Exercise*, 24(7), 755–763.

Clarkson, P., Kearns, A., Rouzier, P., Rubin, R., & Thompson, P. (2006). Serum creatine kinase levels and renal function measures in exertional muscle damage. *Med Sci Sports Exercise*, 38(4), 623–627.

Cogan, K. D. (2006). Sport psychology in gymnastics. In J. Dosil (ed.). *The Sport Psychologist's Handbook* (pp. 641–661). Chichester, UK: Wiley.

Cogan, K. D. & Vidmar, P. (2000). *Sport Psychology Library: Gymnastics*. Morgantown, WV. Fitness Information Technology.

Corbin, C. (1972). Mental practice. In W. Morgan (ed.), *Ergogenic Aids and Muscular Performance* (pp. 688–784). New York: Academic Press.

Cormie, P., Sands, W. A., & Smith, S. L. (2004). A comparative case study of Roche vaults performed by elite male gymnasts. *Technique*, 24(8), 6–9.

Côté, J. (1999). The influence of the family in the development of talent in sports. *The Sport Psychologist*, 13, 395–417.

Côté, J., Baker, J., & Abernethy, B. (2003). From play to practice: A developmental framework for the acquisition of expertise in team sports. In J. S. Starkes & K. A. Ericsson (eds). *Expert Performance In Sports: Advances in Research on Sport Expertise* (pp. 89–113). Champaign, IL: Human Kinetics.

Côté, J., & Hay, J. (2002). Children's involvement in sport: A developmental perspective. In J. M. Silva & D. Stevens (eds). *Psychological Foundations in Sport* (2nd. ed., pp. 484–502.). Boston: Merrill.

Courteix, D., Lespessailles, E., Obert, P., & Benhamou, C. L. (1999). Skull bone mass deficit in prepubertal highly trained gymnast girls. *Int J Sports Med*, 20(5), 328–333.

Courteix, D., Rieth, N., Thomas, T., Van Praagh, E., Benhamou, C., Collomp, K., *et al.* (2007). Preserved bone health in adolescent elite rhythmic gymnasts despite hypoleptinemia. *Hormone Research*, 68, 20–27.

Crielaard, J. M., & Pirnay, F. (1981). Anaerobic and aerobic power of top athletes. *Eur J Appl Physiol*, 47, 295–300.

Csikszentmihali, M. (1975). *Beyond Boredom and Anxiety*. San Francisco: Jossey-Bass.

Csikszentmihalyi, M., Rathunde, K., & Whalen, S. (1993). *Talented Teenagers: The Roots of Success and Failure*. Cambridge: Cambridge University Press.

Cureton, T. K. (1941). Flexibility as an aspect of physical fitness. *The Research Quarterly*, 12, 381–390.

Davidson, P. L., Mahar, B., Chalmers, D. J., & Wilson, B. D. (2005). Impact modeling of gymnastic backhandsprings and dive-rolls in children. *Journal of Applied Biomechanics*, 21, 115–128.

Davlin, C. D., Sands, W. A., & Shultz, B. B. (2001a). Peripheral vision and back tuck somersaults. *Perceptual and Motor Skills*, 93, 465–471.

Davlin, C. D., Sands, W. A., & Shultz, B. B. (2001b). The role of vision in control of orientation in a back tuck somersault. *Motor Control*, 3, 337–346.

Davlin, C. D., Sands, W. A., & Shultz, B. B. (2002). Influence of vision on kinesthetic awareness while somersaulting. *International Sports Journal*, 6(2), 172–177.

Davlin, C. D., Sands, W. A., & Shultz, B. B. (2004). Do gymnasts 'spot' during a back tuck somersault? *International Sports Journal*, 8(2), 72–79.

Denadai, B. S., Figuera, T. R., Favaro, O. R., & Gonçalves, M. (2004). Effect of the aerobic capacity on the validity of the anaerobic threshold for determination of the maximal lactate steady state in cycling. *Braz J Med Biol Res*, 37(10), 1551–1556.

Dodd, S., Powers, S., Callender, T., & Brooks, E. (1984). Blood lactate disappearance at various intensities of recovery exercise. *J. Appl. Physiol. Resp. Environ. Exercise Physiol*, 57(5), 1462–1465.

Douda, H., & Tokmakidis, S. P. (1997). Muscle strength and flexibility of the lower limbs between rhythmic sports and artistic female gymnasts. Paper presented at the Second Annual Congress of the European College of Sport Science.

Dowthwaite, J. N., & Scerpella, T. A. (2009). Skeletal geometry and indices of bone strength in artistic gymnasts. *J Musculoskelet Neuronal Interact*, 9(4), 198–214.

Drabik, J. (1996). *Children & Sports Training*. Island Pond, VT: Stadion Publishing Co.

Drinkwater, B., Nison, K., Chesnut, C. H., Bremmer, W. J., Shainholtz, S., & Southworth, M. B. (1984). Bone mineral content of amenorrheic and eumenorrheic athletes. *N Engl J Med*, 311(5), 277–281.

Driss, T., Vandewalle, H., & Monod, H. (1998). Maximal power and force–velocity relationships during cycling and cranking exercises in volleyball players: Correlation with the vertical jump test. *J Sports Med Phys Fitness*, 38(4), 286–293.

Ducher, G., Hill, B. L., Angeli, T., Bass, S. L., & Eser, P. (2009). Comparison of pQCT parameters between ulna and radius in retired elite gymnasts: The skeletal benefits associated with long-term gymnastics are bone- and site-specific. *J Musculoskelet Neuronal Interact*, 9(4), 247–255.

Durand-Bush, N. (1995). Validity and reliability of the Ottawa Mental Skills Assessment Tool (OMSAT–3). Unpublished master's thesis, School of Human Kinetics, University of Ottawa.

Durand-Bush, N., & Salmela, J. H. (2002). The development and maintenance of expert athletic performance: Perceptions of world and Olympic champions. *Journal of Applied Sport Psychology*, 14, 154–171.

Durand-Bush, N., & Salmela, J.H. ( 2001). The development of talent in sport. In R. N. Singer, H.a., & C. M. Janelle (Eds.) *Handbook of Sport Psychology* (2nd ed.) (pp. 269–289) New York: Wiley.

Durand-Bush, N., Salmela, J. H., & Green-Demers, I. (2001). The Ottawa Mental Skills Assessment Tool (OMSAT-3*). *The Sport Psychologist*, 15, 1–19.

Easterbrook, J. A. (1959). The effect of emotion on cue utilization and the organization of behavior. *Psychological Review*, 66, 183–201.

Ericsson, K. A., (2007). Deliberate practice and the modifiability of body and mind: toward a science of the structure and acquisition of expert and elite performance. *International Journal of Sport Psychology*, 38, 1, 4–34.

Ericsson, K. A., Krampe, R. T., & Tesch-Römer, C. (1993). The role of deliberate practice in the acquisition of expert performance. *Psychological Review*, 100, 363–406.

Falgairette, G., Bedu, M., Fellmann, N., Van Praagh, E., & Coudert, J. (1991). Bio-energetic profile in 144 boys aged from 6 to 15 years with special reference to sexual maturation. *Eur J Appl Physiol*, 62, 151–156.

Faria, I. E., & Faria, E. W. (1989). Relationship of the anthropometric and physical characteristics of male junior gymnasts to performance. *J Sports Med Phys Fitness*, 29(4), 369–378.

Faria, I. E., & Phillips, A. (1970). A study of telemetered cardiac response of young boys and girls during gymnastic participation. *J Sports Med*, 10, 145–160.

Fédération Internationale de Gymnastique (FIG) (2009). *Code of Points for Men's Artistic Gymnastics Competitions (3rd)*. Lausanne: Fédération Internationale de Gymnastique.

Feigley, D. A. (1987). Coping with fear in high level gymnastics. In J. H. Salmela, B. Petiot & T. B. Hoshizaki (eds), *Psychological Nurturing and Guidance of Gymnastic Talent* (pp. 13–27). Montreal: Sport Psyche.

Feynman, R. (1965). *The Character of Physical Law*. Cambridge, MA: MIT Press.

Filaire, E., & Lac, C. (2002). Nutritional status and body composition of juvenile elite female gymnasts. *J Sports Med Phys Fitness*, 42(1), 65–70.

Fitts, M. A. (1995). Serum CPK reported in LSAH participants. *The Longitudinal Study of Astronaut Health Newsletter*, 4, 5–6.

Flores, J. O. (2001). Lactate dehydrogenase isoenzymes test. *Encyclopedia of Medicine* http://findarticles.com/p/articles/mi_g2601/is_0008/ai_2601000803

Fogarty, G. J. (1995). Some comments on the use of psychological tests in sport settings. *International Journal of Sport Psychology*, 26, 161–170.

Fogelholm, G. M., Kukkonen–Harjula, T. K., Taipale, S. A., Sievänen, H. T., Oja, P., & Vuori, I. (1995). Resting metabolic rate and energy intake in female gymnasts, figure-skaters and soccer players. *Int J Sports Med*, 16(8), 551–556.

Fournier, J., Calmels, C., Durand-Bush, N., & Salmela, J. H. (2005). Effects of a season-long PST program on gymnastic performance and on psychological skill development. *ISJEP*, 1, 7–25.

French, D. N., Gómez, A. L., Volek, J. S., Rubin, M. R., Ratamess, N. A., Sharman, M. J., *et al.* (2004). Longitudinal tracking of muscular power changes of NCAA division I collegiate women gymnasts. *J Strength Cond Res*, 18(1), 101–107.

Fry, A. C., Ciroslan, D., Fry, M. D., LeRoux, C. D., Schilling, B. K., & Chiu, L. Z. (2006). Anthropometric and performance variables discriminating elite American junior men weightlifters. *J Strength Cond Res*, 20(4), 861–866.

Gabbett, T. J. (2006). Performance changes following a field conditioning program in junior and senior rugby league players. *J Strength Cond Res*, 20(1), 215–221.

Garnier, P., Mercier, B., Mercier, J., Anselme, F., & Préfaut, C. (1995). Aerobic and anaerobic contribution to Wingate test performance in sprint and middle-distance runners. *Eur J Appl Physiol*, 70, 58–65.

Gateva, M., & Andonov, K. (2005). Updating the system of evaluation and control of the physical preparation in rhythmic gymnastics. *Sports and Science*, Special Edition, 2, 128–137.

Georgopoulos, N. A., Markou, K. B., Theodoropoulou, A., Bernardot, D., Leglise, M., & Vagenakis, A. G. (2002). Growth retardation in artistic compared with rhythmic elite female gymnasts. *J Clin Endocrinol Metab*, 87(7), 3169–3173.

Goh, T. C., & Fock, K. M. (1985). Elevation of creatine phosphokinase in heat syndrome. *Singapore Medical Journal*, 26(4–5), 369–371.

Goswami, A., & Gupta, S. (1998). Cardiovascular stress and lactate formation during gymnastic routines. *J Sports Med Physical Fitness*, 38, 317–322.

Gould, D. (1998). Goal-setting for peak performance. In J. Williams (ed.). *Personal Growth to Peak Performance* (2nd ed., pp. 182–196) Mountain View, CA: Mayfield.

Gould, D., Guinan, D., Greenleaf, C., Medbery, R., Strickland , M., Lauer, L., *et al.* (1998). *Positive and Negative Factors Influencing U.S. Olympic Athletes and Coaches: Atlanta Games Assessment*. Colorado Springs, CO: U.S. Olympic Committee.

Gould, D., & Krane, V. (1993). The arousal–athletic relationship: Current status and future directions. In T. S. Horn (ed.), *Advances in Sport Psychology* (pp. 119–141). Champaign, IL: Human Kinetics.

Haguenauer, M., Legreneur, P., & Monteil, K. M. (2005). Vertical jumping reorganization with aging: A kinematic comparison between young and elderly men. *J Appl, Biomech*, 21(3), 236–246.

Hanin, Y., & Hanina, M. (2009). Optimization of performance in top-level athletes: An action-focused coping approach. *International Journal of Sports Science & Coaching*, 4(1), 47–55.

Hardy, L. (1990). A catastrophe model of performance in sport. In J. G. Jones & L. Hardy (eds), *Stress and Performance in Sport* (pp. 81–106). Chichester, UK: John Wiley & Sons.

Harre, D. (1982). *Principles of Sports Training*. Berlin: Sportverlag.

Harris, D. V., & Williams, J. M. (1993). Relaxation and energizing techniques for regulation of arousal. In J. M. Williams (ed.). *Applied Sport Psychology: Personal Growth to Peak Performance* (2nd ed., pp. 185–199). Mountain View, CA: Mayfield.

Hars, M., Holvoet, P., Barbier, F., Gillet, C., & Lepoutre, F. X. (2008). Study of impulses during a walkover backward on the balance beam in women gymnasts. Paper presented at the 1st scientific symposium of the Asian Gymnastics Union.

Hay, J. G. (1973). *The Biomechanics of Sports Techniques*. Englewood Cliffs, NJ: Prentice Hall.

Hellebrandt, F. A., Parrish, A. M., & Houtz, S. J. (1947). Cross education. *Archives of Physical Medicine and Rehabilitation*, 28, 76–85.

Henry, F. M., & Rogers, D. E. (1960). Increased response latency for complicated movements and a 'Memory Drum' theory of neuromotor reaction. *The Research Quarterly*, 31(3), 448–458.

Hickson, R. C., Dvorack, B. A., Gorostiaga, E. M., Kurowski, T. T., & Foster, C. (1988). Potential for strength and endurance training to amplify endurance performance. *J Appl Physiol*, 65, 2285–2290.

Hiley, M. J., & Yeadon, M. R. (2008). Optimisation of high bar circling technique for consistent performance of a triple piked somersault dismount. *Journal of Biomechanics*, 41, 1730–1735.

Hodgkins, J. (1963). Reaction time and speed of movement in males and females of various ages. *The Research Quarterly*, 34(3), 336–345.

Hoeger, W. W. K., & Fisher, G. A. (1981). Energy costs for men's gymnastic routines. *International Gymnast*, 23(1), TS1–TS3.

Holt, J., Holt, L. E., & Pelham, T. W. (1995). Flexibility redefined. Paper presented at the XIII International Symposium on Biomechanics in Sports. International Society of Biomechanics in Sports.

Holvoet, P., Lacouture, P. & Duboy, J. (2002a). Practical use of airborne simulation in a release–regrasp skill on the high bar. *J Appl Biomech*, 18, 332–344.

Holvoet, P., Lacouture, P., Duboy, J., Junqua A. & Bessonnet G. (2002b). Joint forces and moments involved in giant swings on the high bar. *Science & Sports*, 17, 26–30.
Horak, J. (1969). The performance of top sportsmen. *Teor. Praxe. Teel. Vych.*, 16, 18–20.
Horswill, C. A., Miller, J. E., Scott, J. R., Smith, C. M., Welk, G., & Van Handel, P. (1992). Anaerobic and aerobic power in arms and legs of elite senior wrestlers. *Int J Sports Med*, 13, 558–561.
Inbar, O., & Bar-Or, O. (1977). Relationships of anaerobic and aerobic arm and leg capacities to swimming performance of 8–12 years old children. In R. J. Shephard & H. Lavallée (eds), *Frontiers of Physical Activities and Child Health* (pp. 238–292). Québec: Du Pélican.
Irwin, D. (1993). *Behind the bench: Coaches talk about life in the NHL*. Toronto: McClelland & Stewart.
Jackson, A. S., Beard, E. F., Wier, L. T., Ross, R. M., Stuteville, J. E., & Blair, S. N. (1995). Changes in aerobic power of men, ages 25–70 yr. *Med Sci Sports Exerc*, 27(1), 113–120.
Jackson, S. A. & Csikszentmihalyi, M. (1999). *Flow in Sports: The Keys to Optimal Experiences and Performances*. Champaign, IL: Human Kinetics.
Jackson, A. S., Wier, L. T., Ayers, G. W., Beard, E. F., Stuteville, J. E., & Blair, S. N. (1996). Changes in aerobic power of women, ages 20–64 yr. *Med Sci Sports Exerc*, 28(7), 884–891.
Jacobson, E. (1938). *Progressive relaxation*. Chicago: University of Chicago Press.
Jancarik, A., & Salmela, J. H (1987). Longitudinal changes in the physiological, organic and perceptual factors in Canadian male gymnasts. In B. Petiot, J. H. Salmela, & T. B. Hoshizaki (Eds.). World identification systems for gymnastic talent (pp. 151–156). Montreal: Sport Psyche.
Jankauskienė, R., & Kardelis, K. (2005). Body image and weight reduction attempts among adolescent girls involved in physical activity. *Medicina* (Kaunas), 41(9), 796–801.
Jansson, E., Sjodin, B., & Tesch, P. (1978). Changes in muscle fiber type distribution in man after physical training. A sign of fiber type transformation? *Acta Physiol Scand*, 104, 235–237.
Jemni, M. (2001). *Etude du profil bioénergétique et de la récupération chez des gymnastes*. Université Rennes 2 Haute Bretagne, Rennes – France.
Jemni, M. (2008) *Proceedings of the Scientific Seminar 'The latest research in artistic gymnastics'. Held in parallel to the 4th Asian Artistic Gymnastics Games*. Asian Gymnastics Union. Aspire Academy, Doha, Qatar, 14th Nov 2008.
Jemni, M. (2010) Recovery modalities: Effects on hormones' balance and performance. Delivered at 5th International Scientific Congress 'Sport, Stress, Adaptation'. National Sport Academy Vasil Levski, Sofia, Bulgaria. 23–25 April 2010.
Jemni, M., Friemel, F., Le Chevalier, J. M., & Origas, M. (1998). Bioénergétique de la gymnastique de haut niveau. *Education Physique et Sportive*. 39, 29–34.
Jemni, M., Friemel, F., Le Chevalier, J. M., & Origas, M. (2000). Heart rate and blood lactate concentration analysis during a high level men's gymnastics competition. *J Strength Cond Res*, 14(4), 389–394.
Jemni, M., Friemel, F., & Sands, W. (2002). Etude de la récupération entre les agrès lors de quatre séances d'entraînement de gymnastique masculine. *Education Physique et Sportive*. 57, 57–61.
Jemni, M., Friemel, F., Sands, W. A., & Mikesky, A. (2001). Evolution of the physiological profile of gymnasts over the past 40 years. (Review). *Can J Appl Physiol*, 26(5), 442–456.
Jemni, M., Keiller, D., & Sands, W. A. (2008). Are there any health risks associated to high training loads in highly trained gymnasts? Paper presented at the 13th Annual Congress of the European College of Sport Science (ECSS).
Jemni, M., & Robin, J. F. (2005, 11–13 April). *Proceedings of the 5th International Conference of the AFRAGA (Association Française de Recherche en Activités Gymniques et Acrobatiques)*. Hammamet, Tunisia.
Jemni, M., & Sands, W. (2000). La planification de l'entraînement en gymnastique. Exemple : la dernière semaine avant la compétition. *Gym technic* 31, 17–20.
Jemni, M., & Sands, W. (2003). Heart rate and blood lactate concentration as training indices for high-level men's gymnasts. *Elite Gymnastics Journal*. 26(4), 18–23.
Jemni, M., Sands, W. A., Friemel, F., Cooke, C., & Stone, M. (2006). Effect of gymnastics training on aerobic and anaerobic components in elite and sub elite men gymnasts. *J Strength Cond Res*, 20(4), 899–907.
Jemni, M., Sands, W., Friemel, F., & Delamarche, P. (2003). Effect of active and passive recovery on blood lactate and performance during simulated competition in high level gymnasts. *Can. J. Appl. Physiol*. 28(2), 240–256.

Jerome, W., Weese, R., Plyley, M., Klavora, P., & Howley, T. (1987). The Seneca gymnastics experience. In J. H. Salmela, B. Petiot, & T. B. Hoshizaki (eds). *Psychological Nurturing and Guidance of Gymnastic Talent.* (pp. 90–119) Montreal: Sport Psyche.

Joch, W. (1990). Dimensions of motor speed. *Modern Athlete and Coach*, 28(2), 25–29.

Jones, G., Hanton, S., & Connaughton, D. (2002). What is this thing called mental toughness? An investigation of elite sport performers. *Journal of Applied Sport Psychology*, 14, 205–218.

Jones, G., & Swain, A. (1995). Predisposition debilitative and facilitative anxiety in elite and nonelite performers. *The Sport Psychologist*, 9, 201–211.

Jurimae, J., & Abernethy, P. J. (1997). The use of isoinertial, isometric and isokinetic dynamometry to discriminate between resistance and endurance athletes. *Biology of Sport*, 14(2), 163–171.

Kaman, R. L., Goheen, B., Patton, R., & Raven, P. (1977). The effects of near maximum exercise on serum enzymes. The exercise profile versus the cardiac profile. *Clin Chim Acta*, 81, 145–152.

Kerr, J. H. (1997). *Motivation and Emotion in Sport: Reversal Theory*. East Sussex, UK: Psychology Press.

King, M. A., & Yeadon, M. R. (2004). Maximizing somersault rotation in tumbling. *Journal of Biomechanics*, 37, 471–477.

King, M. A., & Yeadon, M. R. (2005). Factors influencing performance in the Hecht vault and implications for modelling. *Journal of Biomechanics*, 38, 145–151.

Kirby, R. L., Simms, F. C., Symington, V. J., & Garner, G. B. (1981). Flexibility and musculoskeletal symptomatology in female gymnasts and age-matched controls. *Am J Sports Med*, 9, 160–164.

Know, Y. H., Fortney, V. L., & Shin, I. S. (1990). Analysis of Yurchenko vaults performed by female gymnasts during the 1988 Seoul Olympic Games. *Int J Sport Biomech*, 2(6), 157–177.

Knuttgen, H. G., & Komi, P. V. (1992). Basic definitions for exercise. In P. V. Komi (ed.), *Strength and Power in Sport* (pp. 3–6). Oxford, England: Blackwell Scientific Publications.

Koh, M., & Jennings, L. (2007). Strategies in preflight for an optimal Yurchenko Layout Vault. *Journal of Biomechanics*, 40, 471–477.

Krestovnikov, A. N. (1951). *Ocerki po fisiologii fiziceskich upraznenij*. Moskova: FIS.

Kruse D, Lemmen B. (2009). Spine injuries in the sport of gymnastics. Curr Sports Med Rep. 8(1):20–28.

La Porta, M. A., Linde, H. W., Bruce, D. L., & Fitzsimons, E. J. (1978). Elevation of creatine phosphokinase in young men after recreational execise. JAMA, 239, 2685–2686.

Lazurus, R., & Folkman, S. (1984). *Stress, Appraisal and Coping*. New York: Springer.

Lechevalier, J. M., Origas, M., Stein, J. F., Fraisse, F., Barbierie, L., Mermet, P., *et al.* (1999). Comparaison de 3 séances d'entraînement-type chez des gymnastes espoirs: Confrontation avec les valeurs du métabolisme enregistrées en laboratoire. *Gym Technic*, 27, 24–31.

Lidor, R., Hershko, Y., Bilkevitz, A., Arnon, M., & Falk, B. (2007). Measurement of talent in volleyball: 15-month follow-up of elite adolescent players. *J Sports Med Phys Fitness*, 47(2), 159–168.

Lindholm, C., Hagenfeldt, K., & Hagman, U. (1995). A nutrition study in juvenile elite gymnasts. *Acta Paediatr*, 84(3), 273–277.

Linge, S., Hallingstad, O., & Solberg, F. (2006). Modeling the parallel bars in men's artistic gymnastics. *Human Movement Science*, 25, 221–237.

Loehr, J. E. (1983). The ideal performance state. *Science Periodical on Research and Technology in Sport*, 1–8.

Loko, J., Aule, R., Sikkut, T., Ereline, J., & Viru, A. (2000). Motor performance status in 10- to 17-year-old Estonian girls. *Scand J Med Sci Sports*, 10(2), 109–113.

Loucks, A. B., & Redman, L. M. (2004). The effect of stress on menstrual function. *Trends Endocrinol Metab*, 15, 466–471.

Mahoney, M. J. (1989). Psychological predictors of elite and non-elite performance in Olympic weightlifting. *Int J Sport Psy*, 20, 1–12.

Mahoney, M. J., & Avener, M. (1977). Psychology of the elite athlete: An exploratory study. *Cognitive Therapy and Research*, 3, 361–366.

Mahoney, M. J. (1979). Cognitive skills and athletic performance. In P. C. Kendall & S.D. Hollon (Eds.), Cognitive-behavioral interventions: Theory, research, and procedures. New York: Academic.

Mahoney, M. J., Gabriel, T. J., & Perkins, T. S. (1987). Psychological skills and exceptional athletic performance. *The Sport Psychologist*, 1, 189–199.

Mahoney, M. J. (1989). Psychological predictors of elite and non-elite performance in Olympic weightlifting. *International Journal of Sport Psychology*, 20, 1–12.

Malina, R. M. (1994). Physical activity and training: Effects on stature and the adolescent growth spurt. *Medicine & Science in Sports & Exercise*, 26(6), 759–766.

Malina, R. M. (1996). Growth and maturation of female gymnasts. *Spotlight on Youth Sports*, 19(3), 1–3.

Marcinik, E. J., Potts, J., Scholabach, G., Will, S., Dawson, P., & Hurley, B. F. (1991). Effects of strength training on lactate threshold and endurance performance. *Med Sci Sports Exerc*, 23(6), 739–743.

Marina, M., Jemni, M., & Rodríguez, F. A. (2011). Plyometric jumping performances of elite male and female gymnasts. *J Strength Cond Res* (in press).

Markou, K. B., Mylonas, P., Theodoropoulou, A., Kontogiannis, A., Leglise, M., Vagenakis, A. G., *et al.* (2004). The influence of intensive physical exercise on bone acquisition in adolescent elite female and male artistic gymnasts. *J Clin Endocrinol Metab*, 89(9), 4383–4387.

Markovic, G., Dizdar, D., Jukic, I., & Cardinale, M. (2004). Reliability and factorial validity of squat and countermovement jump tests. *J Strength Cond Res*, 18(3), 551–555.

Martens, R. (1977). *Sport Competition Anxiety Test*. Champaign, IL Human Kinetics.

Martin, K. A. (2002). Development and validation of the coaching staff cohesion scale. *Measurement in Physical Education and Exercise Science*, 6(1), 23–42.

Matejek, N., Weimann, E., Witze, I. C., Mölenkamp, S., Schwidergall, S., & Böhles, H. (1999). Hypoleptinemia in patients with anorexia nervosa and in elite gymnasts with anorexia athletica. *Int J Sports Med*, 20, 451–456.

Matveyev, L. (1977). *Fundamentals of Sports Training*. Moscow, USSR: Progress Publishers.

Mayhew, J. L., & Salm, P. C. (1990). Gender differences in anaerobic power tests. *European Journal of Applied Physiology and Occupational Physiology*, 60, 133–138.

McArdle, W., Katch, F., & Katch, V. (2005). *Essentials of Exercise Physiology* ( 3rd ed.). PA: Lippincott Williams & Wilkins.

Melrose, D. R., Spaniol, F. J., Bohling, M. E., & Bonnette, R. A. (2007). Physiological and performance characteristics of adolescent club volleyball players. *J Strength Cond Res*, 21(2), 481–486.

Mero, A. (1998). Power and speed training during childhood. In V. P. E. (ed.), *Pediatric Anaerobic Performance* (pp. 241–267). Champaign, IL: Human Kinetics.

Mikulas, S. (1994). Evolution du niveau de l'état fonctionnel de l'analyseur vestibulaire en gymnastique sportive (garçons). In M. Ganzin (ed.), *Gymnastique artistique et GRS. Communications scientifiques et techniques d'experts étrangers*. Paris, France: INSEP.

Milev, N. (1994). Analyse cinématique comparative du double salto arrière tendu avec et sans vrille (360°) à la barre fixe. In M. Ganzin (ed.), *Gymnastique artistique et GRS. Communications scientifiques et techniques d'experts étrangers*. Paris, France: INSEP.

Mills, M., Pain, M., & Yeadon, M. R. (2006). Modeling a viscoelastic gymnastics landing mat during impact. *Journal of Applied Biomechanics*, 22, 103–111.

Mkaouer, B., Jemni, M., Amara, S. M., Abahnini, K., Agrebi, B., Tabka, Z., *et al.* (2005). Analyse des paramètres déterminants de la performance lors du grand jeté lancer-rattraper en GR. Paper presented at the 5th International Conference of the Association Française pour la Recherche en Activités Gymniques et Acrobatiques (AFRAGA).

Mkaouer, B., Jemni, M., Amara, S., & Tabka, Z. (2008). Kinematics study of jump in backward rotation. Paper presented at the 1st Scientific Symposium of the Asian Gymnastics Union.

Moffroid, M., & Whipple, R. H. (1970). Specificity of speed of exercise. *Physical Therapy*, 50, 1693–1699.

Montgomery, D. L., & Beaudin, P. A. (1982). Blood lactate and heart rate response of young females during gymnastic routines. *J Sports Medicine*, 22, 358–364.

Montpetit, R. (1976). Physiology of gymnastics. In J. Salmela (ed.), *The Advanced Study of Gymnastics*: Springfield, Il.: C.C. Thomas Publisher.

Montpetit, R., & Matte, G. (1969). Réponses cardiaques durant l'exercice de gymnastique. *Kinanthropologie*, 1, 211–222.

Moraes, L. C., Salmela, J. H., Rabelo, A. S., & Vianna, Jr., N. S. (2004). Le développement des jeunes joueurs braziliens au football et au tennis: Le role des parents. *STAPS*, 64, 108–126.

Moritani, T., & DeVries, H. A. (1979). Neural factors versus hypertrophy in the time course of muscle strength gain. *American Journal of Physical Medicine & Rehabilitation*, 58(3), 115–130.

Moschos, S., Chen, J. L., & Mantzoros, C. S. (2002). Leptin and reproduction; a review. *Fertil Steril*, 77, 433–444.

Müller, E., Raschner, C., & Schwameder, H. (1999). The demand profile of modern high-performance training. In F. L. E. Müller, & G. Zallinger, (eds), *Science in Elite Sport* (pp. 1–31). London, England: E & FN Spon.

Muñoz, M. T., de la Piedra, C., Barrios, V., Garrido, G., & Argente, J. (2004). Changes in bone density and bone markers in rhythmic gymnasts and ballet dancers: Implications for puberty and leptin levels. *Eur J Endocrinol*, 151(4), 491–496.
Murphy, S. M., & Jowdy, D. P. (1993). Imagery and mental practice. In T. S. Horn (ed.), *Advances In Sport Psychology* (pp. 221–250). Champaign, IL: Human Kinetics.
Murray, J. (1989). An investigation of competitive anxiety versus positive affect. Unpublished master's thesis, University of Virginia, Charlottesville, VA.
Nideffer, R. M. (1988). Issues in the use of psychological tests in applied settings. *The Sport Psychologist*, 1, 18–28.
Niedeffer, R. M. (1976). Test of attentional and interpersonal style. *Journal of Personality and Social Psychology*, 34, 394–404.
Niemi, M. B. (2009). Cure in the mind. *Scientific American Mind*. (February/March), 42–49.
Noble, L. (1975). Heart rate and predicted $VO_2$ during women's competitive gymnastic routines. *Journal of Sports Medicine & Physical Fitness*, 15(2), 151–157.
Obert, P., Stecken, F., Courteix, D., Germain, P., Lecoq, A. M., & Guenon, P. (1997). Adaptations myocardiques chez l'enfant prépubère soumis à un entraînement intensif. Etude comparative entre une population de gymnastes et de nageurs. *Science et Sports*, 12, 223–231.
Oda, S., & Moritani, T. (1994). Maximal isometric force and neural activity during bilateral and unilateral elbow flexion in humans. *European Journal of Applied Physiology and Occupational Physiology*, 69, 240–243.
Ogilvie, B.C., & Tutko, T. A. (1966). *Problem Athletes and How to Handle Them*. London: Pelham Books
Olbrecht, J. (2000). *The Science Of Winning*. Luton, England: Swimshop.
Oliver, L. R., De Waal, A., & Retief, F. J. (1978). Electrocardiographic and biochemical studies on marathon runners. *S Afr Med J*, 53, 783–787.
Orlick, T. (2008). *In Pursuit of Excellence* (4th ed.) Champaign, IL: Human Kinetics.
Orlick, T., & Partington, J. (1988). *Psyched: Inner Views of Winning*. Ottawa, ON, Canada: The Coaching Association of Canada.
Oudejans, R. R. D. (2008). Reality-based practice improves handgun shooting performance of police officers. *Ergonomics*, 81 (3), 261–273.
Perel, E., & Killinger, D. W. (1979). The interconversion and aromatization of androgens by human adipose tissue. *J Steroid Biochem*, 10, 623–627.
Petrov, V. (1994a). Modèle expérimental d'exécution du double salto avant groupe avec reprise de barre à la barre fixe. In M. Ganzin (ed.), *Gymnastique artistique et GRS. Communications scientifiques et techniques d'experts étrangers* (pp. 135–142). Paris, France: INSEP.
Petrov, V. (1994b). Technique et méthode d'exécution d'un salto avant jambes écartées à partir d'un grand tour jusqu'à la reprise de la barre en suspension arrière. In M. Ganzin (ed.), *Gymnastique artistique et GRS. Communications scientifiques et techniques d'experts étrangers* (pp. 169–175). Paris, France: INSEP.
Pool, J., Binkhorst, R. A., & Vos, J. A. (1969). Some anthropometric and physiological data in relation to performance of top female gymnasts. *Internationale Zeitschrift Für Angewandte Physiologie*, 27, 329–338.
Potiron-Josse, M., & Bourdon, A. (1989). Le gros cœur du sportif. *Science & Sports*, 4(4), 305–316.
Prassas, S. G. (1988). Biomechanical model of the press handstand in gymnastics. *International Journal of Sport Biomechanics*, 4(4), 326–341.
Rabelo, A. S. (2002). The role of families in the development of aspiring expert soccer players. Unpublished master's thesis, Federal University of Minas Gerais, Brazil.
Rabelo, A. S., Moraes, L.C., & Salmela, J.H. (2001). The role of parents in the development of young Brazilian athletes in soccer. *Conference Proceedings of the Association for the Advancement of Applied Sport Psychology*. Orlando, FL: pp. 52–53.
Ravizza, K. H. (2002). A philosophical construct: A framework for performance enhancement. *International Journal of Sport Psychology*, 33, 4–18.
Ravizza, K., & Rotella, R. (1982). Cognitive somatic behavioral interventions in gymnastics. In L. Zaichkowsky & W. E. Sime (eds), *Stress Management for Sport* (pp. 25–35). Reston, VA: AAHPERD.
Régnier, G., & Salmela, J. H. (1987). Predictors of success in Canadian male gymnasts. In J. H. S. B. Petiot & H. B. Hoshizaki (eds), *World Identification Systems for Gymnastic Talent* (pp. 141–150). Montreal: Sport Psyche.
Richards, J. E., Ackland, T. R., & Elliott, B. C. (1999). The effect of training volume and growth on gymnastic performance in young women. *Pediatric Exercise Science*, 11(4), 349–363.

Riley, W. J., Pyke, F. S., Roberts, A. D., & England, J. F. (1975). The effect of long-distance running on some biochemical variables. *Clin Chim Acta*, 65(83–89).

Rodríguez, F. A., Marina, M., & Boucharin, E. (1999). Physiological demands of women's competitive gymnastic routines. Paper presented at the 4th Annual Congress of the European College of Sport Science.

Rosen, L. W., & Hough, D. O. (1988). Pathogenic weight-control behaviors of female college gymnasts. *Physician & Sportsmedicine*, 16(9), 140–143;146.

Roskamm, H. (1980). Le système de transport de l'oxygène. In P. O. Astrand & K. Rodahl (eds), *Précis de physiologie de l'exercice musculaire* (2nd ed., pp. 316–317). Paris: Masson.

Rotella, R. J., & Lerner, J. D. (1993). Responding to competitive pressure. In R. N. Singer, M. Murphey, & L. K. Tennant (eds), *Handbook of Research on Sport Psychology* (pp. 528–541). New York: Macmillan.

Sale, D. G. (1986). Neural adaptation in strength and power training. In N. L. Jones, N. M. McCartney, & A. J. McComas (eds), *Human Muscle Power* (pp. 289–308). Champaign, IL: Human Kinetics.

Sale, D. G. (1992). Neural adaptation to strength training. In P. V. Komi (ed.), *Strength and Power in Sport* (pp. 249–265). Oxford, England: Blackwell Scientific Publications.

Sale, D., & MacDougall, D. (1981). Specificity in strength training: A review for the coach and athlete. *Canadian Journal of Applied Sport Sciences*, 6(2), 87–92.

Sale, D. G., & Norman, R. W. (1982). Testing strength and power. In J. D. MacDougall, H. A. Wenger & H. J. Green (eds), *Physiological Testing of the Elite Athlete* (pp. 7–38). Ithaca, NY: Movement Publications.

Salmela, J. H. (1976). *The Advanced Study of Gymnastics*. Springfield, IL: C.C. Thomas.

Salmela, J. H. (1976). Psychomotor task demands of gymnastics. In J. H. Salmela (ed.), *The Advanced Study of Gymnastics* (pp. 5–19). Springfield, Il: C. C. Thomas.

Salmela, J. H. (1989). Long term intervention with the Canadian men's gymnastics team. *The Sport Psychologist*, 3, 340–349.

Salmela, J. H. (1996). *Great Job Coach!* Ottawa: Potentium.

Salmela, J. H. (2004). Coaching, families and learning in Brazilian youth football players. *Insight*, 7 (2), 36–37.

Salmela, J. H., Marques, M. P., & Machado, R., (2004). The informal structure of football in Brazil. *Insight*, 7(1), 17–19

Salmela, J.H., & Moraes, L.C. (2004). Development of expertise: The role of coaching, families, and cultural contexts. In J. L. Starkes & K.A. Ericsson (Eds.), Expert performance in sports. (pp. 275–194). Champaign, IL: Human Kinetics.

Salmela, J. H., Monfared, S. F., Mosayebi, S. S. & Durand-Bush, N. (2009). Mental skill profiles and expertise levels of elite Iranian athletes. *Int J Sport Psy*, 39, 361–373.

Salmela, J. H., Mosayebi, F., Monfared, S. S., & Durand-Bush, N. (2007). Perceptions of Iranian athletes and coaches of the effectiveness of mental training interventions at the Asian Games. In Y. Theodorakis, M. Goudas, & A. Papaionnou (eds), *Sport and Exercise Psychology: Bridges between Disciplines and Cultures* (pp. 92–96). Thesaloniki: University of Thesaly.

Salmela J. H., Petiot B., Hallé, M., & Régnier, G. (1980). *Competitive Behaviors of Olympic Gymnasts*. Springfield, IL: C. C. Thomas.

Saltin, B., & Astrand, P. O. (1967). Maximal oxygen uptake in athletes. *J. Appl. Physiol.*, 23, 353–358.

Sands, B., & McNeal, J. R. (1999). Body size and sprinting characteristics of 1998 National TOPs athletes. *Technique*, 19(5), 34–35.

Sands, W. A. (1981). *Beginning Gymnastics*. Chicago, IL: Contemporary Books.

Sands, W. A. (1984). *Coaching Women's Gymnastics*. Champaign, IL: Human Kinetics.

Sands, W. A. (1985). Conditioning for gymnastics: A dilemma. *Technique*, 5(3), 4–7.

Sands, W. A. (1987). The edge of the envelope. *Gymnastics Safety Update*, 2(3), 2–3.

Sands, W. A. (1990a). Determining skill readiness. *Technique*, 10(3), 24–27.

Sands, W. A. (1990b). National women's tracking program. *Technique*, 10, 23–27.

Sands, W. A. (1991a). Monitoring elite gymnastics athletes via rule based computer systems. *Masters of Innovation III* (pp. 92). Northbrook, IL: Zenith Data Systems.

Sands, W. A. (1991b). Monitoring the elite female gymnast. *National Strength and Conditioning Association Journal*, 13(4), 66–71.

Sands, W. A. (1993a). The role of science in sport. *Technique*, 13(10), 17–18.

Sands, W. A. (1993b). *Talent Opportunity Program*. Indianapolis, IN: United States Gymnastics Federation.
Sands, W. A. (1994a). Physical abilities profiles – 1993 National TOPs testing. *Technique*, 14(8), 15–20.
Sands, W. A. (1994b). The role of difficulty in the development of the young gymnast. *Technique*, 14(3), 12–14.
Sands, W. A. (1995). How can coaches use sport science? *Track Coach*, 134 (winter), 4280–4283.
Sands, W. A. (1996). How effective is rescue spotting? *Technique*, 16(9), 14–17.
Sands, W. A. (1998). A look at training models. *Technique*, 19, 6–8.
Sands, W. A. (2000a). Injury prevention in women's gymnastics. *Sports Medicine*, 30(5), 359–373.
Sands, W. A. (2000b). Olympic preparation camps 2000 physical abilities testing. *Technique*, 20, 6–19.
Sands, W. A. (2000c). Physiological aspects of gymnastics. Paper presented at the 2èmes Journées Internationales d'Etude de L'Association Française de Recherche en Activités Gymniques et Acrobatiques (AFRAGA).
Sands, W. A. (2000d). Vault run speeds. *Technique*, 20(4), 5–8.
Sands, W. A. (2003). Physiology. In W. A. Sands, D. J. Caine, & J. Borms (eds), *Scientific Aspects of Women's Gymnastics* (pp. 128–161). Basel, Switzerland: Karger.
Sands, W. A., & Cheetham, P. J. (1986). Velocity of the vault run: Junior elite female gymnasts. *Technique*, 6(3), 10–14.
Sands, W. A., Eisenman, P., Johnson, S., Paulos, L., Abbot, P., Zerkel, S., *et al.* (1987). Getting ready for '88. *Technique*, 7, 12–18.
Sands, W. A., Hofman, M. G., & Nattiv, A. (2002). Menstruation, disordered eating behavior, and stature: A comparison of female gymnasts and their mothers. *International Sports Journal*, 6(1), 1–13.
Sands, W. A., Irvin, R. C., & Major, J. A. (1995). Women's gymnastics: The time course of fitness acquisition. A 1-year study. *Journal of Strength and Conditioning Research*, 9(2), 110–115.
Sands, W. A., Jemni, M., Stone, M., McNeal, J., Smith, S. L., & Piacentini, T. (2005). Kinematics of vault board behaviours – A preliminary comparison. Paper presented at the 5th International Conference of the Association Française pour la Recherche en Activités Gymniques et Acrobatiques (AFRAGA).
Sands, W. A., & McNeal, J. R. (1995). The relationship of vault run speeds and flight duration to score. *Technique*, 15(5), 8–10.
Sands, W. A., & McNeal, J. R. (1999). Judging gymnastics with biomechanics. *Sport Science*, 3(1), sportsci.org/jour/9901/was.html
Sands, W. A., McNeal, J., & Jemni, M. (2000). Should Female Gymnasts Lift Weights? *Internet Sports Science Journal*, http://www.sportsci.org/jour/0003/was.html
Sands, W. A., McNeal, J., & Jemni, M. (2001a). Anaerobic Power Profile: Talent-Selected Female Gymnasts Age 9–12 Years. *Technique*, 21, 5–9.
Sands, W. A., McNeal, J., & Jemni, M. (2001b). Fitness profile comparisons: USA women's junior, senior and Olympic gymnastics teams. *J Strength Cond Res*, 15, 398.
Sands, W. A., McNeal, J. R., Ochi, M. T., Urbanek, T. L., Jemni, M., & Stone, M. H. (2004a). Comparison of the Wingate and Bosco anaerobic tests. *J Strength Cond Res*, 18(4), 810–815.
Sands, W. A., McNeal, J., Stone, M., Russell, E., & Jemni, M. (2006b). Flexibility enhancement with vibration: Acute and long-term. *Med Sc Sports Exer*, 38(4), 720–725.
Sands, W. A., McNeal, J. R., Stone, M. H., Smith, S. L., Dunlavy, J. K., Jemni, M., *et al.* (2006a). Exploratory Relationship of Drop Jump Performance with Gymnastics Vaulting and Floor Exercise Scores. Paper presented at the 11th Annual Congress of the ECSS.
Sands, W. A., McNeal, J. R., & Urbanek, T. (2003). On the role of 'Functional Training' in gymnastics and sports. *Technique*, 23(4), 12–13.
Sands, W. A., Smith, S. L., Westenburg, T. M., McNeal, J. R., & Salo, H. (2004b). Kinematic and kinetic case comparison of a dangerous and superior flyaway dismount – women's uneven bars. In M. Hubbard, R. D. Metha, & J. M. Pallis (eds), *The Engineering of Sport 5* (pp. 414–420). Sheffield, UK: International Sports Engineering Association.
Sands, W. A., & Stone, M. H. (2006). Monitoring the elite athlete. *Olympic Coach*, 17(3), 4–12.
Sands, W. A., Stone, M., McNeal, J., Smith, S., Jemni, M., Dunlavy, J., *et al.* (2006c). A Pilot Study to Measure Force Development during a Simulated Maltese Cross for Gymnastics Still Rings. Paper presented at the XXIV International Symposium on Biomechanics in Sports (ISBS).
Schindler, A. E., Ebert, A., & Friedrich, E. (1972). Conversion of androstenedione to estrone by human fat tissue. *J Clin Endocrinol Metab*, 35, 627.

Schmidt, R. A., & Young, D. E. (1991). Methodology for motor learning: A paradigm for kinematic feedback. *Journal of Motor Behavior*, 23(1), 13–24.

Schwartz, R. (2007). Autoimmune and intravascular hemolytic anemias. In L. Goldman & D. Ausiello (eds), *Cecil Medicine*, Philadelphia, PA: Saunders Elsevier.

Seliger, V., Budka, I., Buchberger, J., Dosoudil, F., Krupova, J., Libra, M., *et al.* (1970). Métabolisme énergétique au cours des exercices de gymnastique. *Kinanthropologie*, 2, 159–169.

Selye, H. (1974). *Stress without Distress*. New York: New American Library.

Shaghlil, N. (1978). La gymnastique et son action sur l'appareil circulatoire et respiratoire. Paper presented at the 1er Colloque médical international de gymnastique.

Sheets, A. L. (2008). Evaluation of a subject-specific female gymnast model and simulation of an uneven parallel bar swing. *Journal of Biomechanics*, 4, 326–341.

Shiffrin, R. M. (1976). Capacity limitations in information processing, attention and memory. In W. K. Estes (Ed.), *Handbook of Learning and Cognitive Processes:Attention and Memory*, Vol. 4. Hillsdale, NJ: Erlbaum, pp. 177–236.

Siff, M. C. (2000). *Supertraining*. Denver, CO: Supertraining Institute.

Singer, R. N. (1988). Psychological testing: What value to coaches and athletes? *International Journal of Sport Psychology*, 19, 87–106.

Sipila, S., Koskinen, S. O., Taaffe, D. R., Takala, T. E., Cheng, S., Rantanen, T., *et al.* (2004). Determinants of lower-body muscle power in early postmenopausal women. *J Am Geriatr Soc*, 52(6), 939–944.

Smith, D. J. (2003). A framework for understanding the training process leading to elite performance. *Sports Medicine*, 33(15), 1103–1126.

Soberlack, P. (2001). A retrospective analysis of the development and motivation of professional ice hockey players. Unpublished master's thesis, Queen's University, Kingston, ON.

Soberlak, P. A., & Côté, J. (2003). The developmental activities of elite ice hockey players. *Journal of Applied Sport Psychology*, 15, 41–49.

Soric, M., Misigoj-Durakovic, M., & Pedisic, Z. (2008). Dietary intake and body composition of prepubescent female aesthetic athletes. *Int J Sport Nutr Exerc Metab*, 18(3), 343–354.

Spielberger, C. D. (1966). *Anxiety and Behavior*. New York: Academic Press.

Stamford, B. A., Weltman, A., Moffat, R., & Sady, S. (1981). Exercise recovery above and below the anaerobic threshold following maximal work. *J Appl Physiol*, 51, 840–844.

Stein, N. (1998). Speed training in sport. In B. Elliott (ed.), *Training in Sport* (pp. 288–349). New York, NY: John Wiley & Sons.

Stone, M. H., Stone, M. E., & Sands, W. A. (2007). *Principles and Practice of Resistance Training*. Champaign, IL: Human Kinetics.

Stone, M. H., Wilson, D., Rozenek, R., & Newton, H. (1984). Anaerobic capacity. *National Strength and Conditioning Association Journal*, 5(6), 63–65.

Stroescu, V., Dragan, J., Simionescu, L., & Stroescu, O. V. (2001). Hormonal and metabolic response in elite female gymnasts undergoing strenuous training and supplementation with SUPRO Brand Isolated Soy Protein. *J Sports Med Phys Fitness*, 41(1), 89–94.

Suinn, R. M. (1993). Imagery. In R. N. Singer, M. Murphey & K.Tennant (eds). *Handbook of Research on Sport Psychology* (pp. 492–510) New York: Macmillan.

Svoboda, E. (2009). Avoiding the big choke. *Scientific American Mind*, February/March, 36–41.

Szogy, A., & Cherebetiu, G. (1971). Capacité aérobie maximum chez les sportifs de performance. *Médecine du sport* 45, 224–234.

Takei, Y. (1998). Three-dimensional analysis of handspring with full turn vault: Deterministic model, coaches' beliefs, and judges' scores. *Journal of Applied Biomechanics*, 14, 190–210.

Tanaka, S. (1987). The Japanese gymnastic golden era between the 1960's and 1970's. In B. Petiot, J. H. Salmela, & T. B. Hoshizaki (eds), *World Identification Systems for Gymnastic Talent* (pp. 45–57). Montreal: Sport Psyche.

Tesch, P. A. (1980). Fatigue pattern in subtypes of human skeletal muscle fibers. *International Journal of Sports Medicine*, 1(2), 79–81.

Theodoropoulou, A., Markou, K. B., Vagenakis, G. A., Bernardot, D., Leglise, M., Kourounis, G., *et al.* (2005). Delayed but normally progressed puberty is more pronounced in artistic compared with rhythmic elite gymnasts due to the intensity of training. *J Clin Endocrinol Metab*, 90(11), 6022–6027.

Trappe, S. W., Costill, D. L., Vukovich, M. D., Jones, J., & Melham, T. (1996). Aging among elite distance runners: A 22-yr longitudinal study. *J Appl Physiol*, 80(1), 285–290.

Uneståhl, L-E. (1975). *Hypnosis in the Seventies*. Orebro: Veja.
Vandewalle, H., Peres, G., Heller, J., Panel, J., & Monod, H. (1987). Force–velocity relationship and maximal power on a cycle ergometer. Correlation with the height of a vertical jump. *Eur J Appl Physiol*, 56, 650–656.
Vandewalle, H., Peres, G., Sourabié, O., Stouvenel, & Monod, H. (1989). Force–velocity relationship and maximal anaerobic power during cranking exercise in young swimmers. *Int J Sports Med*, 13, 439–445.
Vankov, I. (1982). *Examination of Physical Preparation of Rhythmic Gymnastics Athletes and Perfecting the System for Control, Evaluation, and Optimization*. Sofia, Bulgaria: National Sport Academy.
Vankov, I. (1983). *Evaluation System of The Physical Preparation In Rhythmic Gymnastics*. Sofia, Bulgaria: ETSPKFKC.
Van Praagh, E., & Dore, E. (2002). Short-term muscle power during growth and maturation. *Sports Med*, 32(11), 701–728.
Vealey, R. S. (1988). Future directions in psychological skills training. *The Sport Psychologist*, 2, 318–336.
Verkhoshansky, Y. V. (1981). Special strength training. *Soviet Sports Review*, 16(1), 6–10.
Verkhoshansky, Y. V. (1985). *Programming and Organization of Training*. Moscow, U.S.S.R: Fizkultura i Spovt.
Verkhoshansky, Y. V. (1996). Speed training for high level athletes. *New Studies in Athletics*, 11(2–3), 39–49.
Verkhoshansky, Y. V. (1998). Organization of the training process. *New Studies in Athletics*, 13(3), 21–31.
Verkhoshansky, Y. V. (2006). *Special Strength Training: A Practical Manual for Coaches*. Moscow, Russia: Ultimate Athlete Concepts.
Vianna, N. S., Jr (2002). The role of families and coaches in the development of aspiring expert tennis players. Unpublished master's thesis, Federal University of Minas Gerais, Brazil.
Ward, K. A., Roberts, S. A., Adams, J. E., Lanham-New, S., & Mughal, M. Z. (2007). Calcium supplementation and weight bearing physical activity: Do they have a combined effect on the bone density of pre-pubertal children? *Bone*, 41(4), 496–504.
Weimann, E. (2002). Gender-related differences in elite gymnasts: The female athlete triad. *J Appl Phsysiol*, 92(5), 2146–2152.
Weimann, E., Blum, W. F., Witzel, C., Schwidergall, S., & Bohels, H. J. (1999). Hypoleptinemia in female and male elite gymnasts. *European Journal of Clinical Investigation*, 29(10), 853–860.
Weimann, E., Witzel, C., Schwidergall, S., & Bohels, H. J. (2000). Peripubertal perturbations in elite gymnasts caused by sport specific training regimes and inadequate nutritional intake. *Int J Sports Med*, 21.
Weinberg R. S., & Gould D. (1999) *Foundations of Sport and Exercise Psychology*. Champaign, IL: Human Kinetics.
Wilk, K. (1990). Dynamic muscle strength testing. In A. L. R. (ed.), *Muscle Strength Testing* (pp. 123–150). New York, NY: Churchill Livingstone.
Williams, J. M. (1986). Psychological characteristics of peak performance. In J. M. Williams (ed.), *Applied Sport Psychology: Personal Growth to Peak Performance* (pp. 123–132). Palo Alto, CA: Mayfield.
Wilmore, J., & Costill, D. (1999). *Physiology of Sport and Exercise* (2 edn). Champaign, IL: Human Kinetics.
Wilmore, J. H., & Costill, D. L. (2005). *Physiology of Sport and Exercise* (3rd edn). Champaign, IL: Human Kinetics.
Winter, E. M. (2005). Jumping: Power or impulse. *Medicine and Science in Sports and Exercise*, 37, 523.
Woodson, W. E., Tillman, B., & Tillman, P. (1992). *Human Factors Design Handbook*. New York, NY: McGraw-Hill.
Yeadon, M. R. (1993a). The biomechanics of twisting somersaults Part I: Rigid body motions. *Journal of Sports Sciences*, 11, 187–198.
Yeadon, M. R. (1993b). The biomechanics of twisting somersaults Part II: Contact twist. *Journal of Sports Sciences*, 11, 199–208.
Yeadon, M. R. (1993c). The biomechanics of twisting somersaults Part III: Aerial twist. *Journal of Sports Sciences*, 11, 209–218.
Yeadon, M. R. (1993d). The biomechanics of twisting somersaults Part IV: Partitioning performances using the tilt angle. *Journal of Sports Sciences*, 11, 219–225.

Yeadon, M. R., & Brewin, M. A. (2003). Optimised performance of the backward longswing on rings. *Journal of Biomechanics*, 36, 545–552.

Yeadon, M. R., King, M. A., & Hiley, M. J. (2005). Computer simulation of gymnastics skills. Paper presented at the 5th International Conference of the Association Française pour la Recherche en Activités Gymniques et Acrobatiques (AFRAGA).

Yeadon, M. R., & Mikulcik, E. C. (1996). The control of non-twisting somersaults using configuration changes. *Journal of Biomechanics*, 29, 1341–1348.

Yerkes, R. M., & Dodson, J. D. (1908). The relation of strength of stimulus to rapidity of habit formation. *Journal of Comparative Neurology of Psychology*, 18, 459–482.

Young, B. W., & Salmela, J. H. (2002). Perceptions of training and deliberate practice of middle distance runners. *International Journal of Sport Psychology*, 33, 167–181.

Yoshida, T., Udo, M., Chida, M., Ichioka, M., Makiguchi, K., & Yamaguchi, T. (1990). Specificity of physiological adaptation to endurance training in distance runners and competitive walkers. *European Journal of Applied Physiology and Occupational Physiology*, 61(3–4), 197–201.

Zaggelidis, S., Martinidis, K., Zaggelidis, G., & Mitropoulou, T. (2005). Nutritional supplements use in elite gymnasts. Physical training. *Electronic Journals of Martial Arts and Sciences*. Retrieved from http://ejmas.com

Zaichkowsky, L., & Takenaka, K. (1993). Optimizing arousal levels. In R. N. Singer, M. Murphey, & L. K. Tennant (eds), *Handbook of Research on Sport Psychology* (pp. 511–527). New York: Macmillan.

# INDEX

Lightning Source UK Ltd.
Milton Keynes UK
UKOW020756200712

196279UK00007B/12/P